Chemistry of Wine Flavor

ACS SYMPOSIUM SERIES **714**

Chemistry of Wine Flavor

Andrew L. Waterhouse, EDITOR
University of California at Davis

Susan E. Ebeler, EDITOR
University of California at Davis

American Chemical Society, Washington, DC

Library of Congress Cataloging-in-Publication Data

Chemistry of Wine Flavor / Andrew L. Waterhouse, editor, Susan E. Ebeler, editor.

p.   cm.—(ACS symposium series , ISSN 0097-6156 ; 714)

"Developed from a symposium sponsored by the Division of Agricultural and Food Chemistry at the 213th National Meeting of the American Chemical Society, San Francisco, California, April 13–17, 1997"—T.p. verso.

Includes bibliographical references (p.   -   ) and index.

ISBN 0–8412–3592–9

1. Wine and wine making—Chemistry—Congresses.

I. Waterhouse, Andrew Leo. II. Ebeler, Susan E., 1961–   . III. American Chemical Society. Division of Agricultural and Food Chemistry. IV. American Chemical Society. Meeting (213th : 1997 : San Francisco, Calif.) V. Series.

TP544.C45   1998
663′.2—dc21   98–34175
                                                                              CIP

PRINTED IN THE UNITED STATES OF AMERICA

Foreword

The ACS Symposium Series was first published in 1974 to provide a mechanism for publishing symposia quickly in book form. The purpose of the series is to publish timely, comprehensive books developed from ACS-sponsored symposia based on current scientific research. Occasionally, books are developed from symposia sponsored by other organizations when the topic is of keen interest to the chemistry audience.

Before agreeing to publish a book, the proposed table of contents is reviewed for appropriate and comprehensive coverage and for interest to the audience. Some papers may be excluded in order to better focus the book; others may be added to provide comprehensiveness. When appropriate, overview or introductory chapters are added. Drafts of chapters are peer-reviewed prior to final acceptance or rejection, and manuscripts are prepared in camera-ready format.

As a rule, only original research papers and original review papers are included in the volumes. Verbatim reproductions of previously published papers are not accepted.

ACS Books Department

Contents

INDEXES

Preface

The complexity of wine composition has always challenged chemists and, as a result, there have been many meetings to discuss the chemistry and the related flavors. Scientific interest in these flavors has led to an increased understanding of wine chemistry, biochemistry, and sensory perception, and meetings of the American Chemical Society (ACS) have provided an important forum for sharing these discoveries: from a presentation by Andre Tchelistcheff on malolactic fermentation at the 1949 ACS meeting in San Francisco to the most recent Wine Flavor Chemistry Symposium at the 1997 meeting in San Francisco, from which this volume is derived.

The first five chapters of this book focus on the grape derived and varietal flavors of wines. Many of these compounds occur as nonvolatile glycosidic flavor precursors and the separation and analysis of these precursors have been a challenging and active field of research. The isolation and quantification of trace volatiles represent examples of the difficulties faced by flavor chemists as they attempt to characterize varietal flavors with sensory thresholds in the parts per trillion range and lower.

The unique flavors of wines are due not only to grape flavors but also to those formed during the primary yeast fermentation and any secondary bacterial or yeast fermentation that can occur. Many of the factors affecting fermentations-related flavors remain controversial (e.g., spontaneous versus inoculated yeast fermentations) or are still not well understood. The effects of grape composition, seasonal variations, and the identification of odor impact compounds need much more investigation. However, novel enzymatic syntheses are leading to an increased understanding of the pathways by which fermentation flavors are formed. These topics are discussed in Chapters 6–9.

The contribution of polyphenols to the bitter taste and astringent mouthfeel of wine is the focus of Chapters 10–12. The effects of grape growing region, wine processing (filtration and fining), and aging are discussed as they relate to polyphenol composition and taste. Finally, the characterization of aromas related to wine maturation, aging in oak cooperage, the cork stopper, and the role of component interactions on flavor volatility and perception are the focus of the final four chapters of the book.

The authors, whose chapters appear in this book, represent a cross-section of the current generation of international experts in the field of wine flavor chemistry. But like all science, current research in wine chemistry builds on the findings of pioneers in the field. For instance, early studies on wine phenolics by Vernon L. Singleton and on wine and sherry volatiles by A. Dinsmoor Webb have been further developed by other scientists and former colleagues in other parts of the world. Although not all of the early wine chemists are individually named in this book, their contributions were essential for achieving our current state of knowledge. As such, this ACS Wine Flavor Chemistry Symposium represented an exciting mixture of topics, scientific history, and recent discoveries, and this proceedings presents one of the most current collections of research on wine flavor chemistry that is available.

We thank the contributors, Diane Eschenbaum for administrative assistance, and the following individuals and organizations who financially contributed to making the symposium a success: The American Society for Enology and Viticulture, the Department of Viticulture and Enology at the University of California at Davis, The E & J Gallo Winery, the Robert Mondavi Winery, ETS Laboratories, and the ACS Division of Agriculture and Food Chemistry.

ANDREW L. WATERHOUSE
Department of Viticulture and Enology
University of California at Davis
Davis, CA 95616–8749

SUSAN E. EBELER
Department of Viticulture and Enology
University of California at Davis
Davis, CA 95616–8749

Chapter 1

Analysis, Structure, and Reactivity of Labile Terpenoid Aroma Precursors in Riesling Wine

Peter Winterhalter[1], Beate Baderschneider, and Bernd Bonnländer

Institut für Pharmazie und Lebensmittelchemie, Universität Erlangen at Nürnberg, Schuhstrasse 19, D-91052 Erlangen, Germany

This chapter discusses the necessity of elucidating the total structure of aroma-relevant glycoconjugates and describes countercurrent chromatographic techniques which enable a gentle isolation of labile aroma precursors from wine. By using one of these all-liquid chromatographic techniques, i.e. multilayer coil countercurrent chromatography (MLCCC), important glycosidic aroma precursors have been recognized for the first time in Riesling wine. The newly identified compounds include the ß-D-glucose ester of (E)-2,6-dimethyl-6-hydroxyocta-2,7-dienoic acid as well as two ß-D-glucopyranosides of 3-hydroxy-7,8-didehydro-ß-ionol. The role of these glycoconjugates in the formation of important wine aroma volatiles is discussed. In addition, the identification of uncommon glycoconjugates in Riesling wine is reported. These novel wine constituents include 2-phenylethyl-α-D-glucopyranoside, the N-glucoside of 2-ethyl-3-methylmaleimide as well as the ß-D-glucose ester of 10,11-dihydroxy-3,7,11-trimethyl-2,6-dodecadienoic acid.

The presence of acid-labile glycoconjugates of monoterpenoids and C_{13}-norisoprenoids in Riesling wine is well documented (*1-8*). The growing interest in these structures in recent days is mainly due to their role as flavour precursors (*9-16*). Especially during a prolonged storage of wine, the acid-catalyzed degradation of such glycoconjugates is considered to make an important contribution to the typical bouquet of bottle-aged wines (*17,18*).

Reasons for Elucidating the Total Structure of Glycosidic Aroma Precursors

Glycosidic aroma precursors are conveniently isolated from grape juice and wine by selective retention on either C_{18}-reversed phase adsorbent (*19*) or Amberlite XAD-2 (*20*), followed by the desorption of the retained glycosides using ethyl acetate or methanol as the eluting solvent. Once a precursor concentrate has been obtained, two lines of investigations can be pursued. The first rapid approach consists of a HRGC-MS analysis of the aglycon fraction obtained after enzymatic hydrolysis. On this basis, some information about the bound aroma fraction is immediately obtained. This approach, however, does not give absolute proof of glycoconjugation.

---

[1]Current address: Institut für Lebensmittelchemie, Technische Universität Braunschweig, Schleinitzstrasse 20, D-38106 Braunschweig, Germany.

In a recent study (*21*), the formation of artifacts during enzymatic hydrolyses has been reported. High concentrations of fungal-derived hydrolases were found to almost completely oxidize some of the aglycon moieties. Glycosides with homo-allylic glycosidic linkages were found to be particularly susceptible to this oxidation. One example is the hydrolysis of glucoconjugated 3-hydroxy-ß-damas-cone **1**. Upon enzymatic hydrolysis with a fungal-derived enzyme preparation, glucoside **1** did not liberate the intact aglycon **2**, instead oxidized products, i.e. the oxodamascones **3** and **4**, were obtained as cleavage products. This observation emphasizes the need to confirm the structures of the glycoconjugates by isolating and characterizing the individual glycoconjugates.

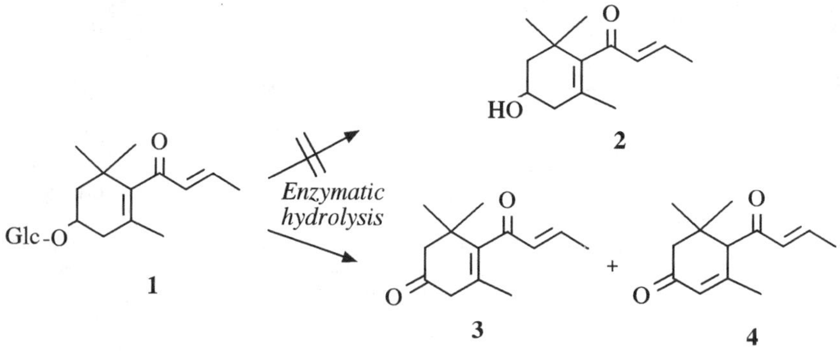

Figure 1. Artifact formation observed after incubation with fungal-derived glycosidase preparations (*21*).

Another reason for elucidating the total structure of the aroma precursor is due to the fact that many of the aroma-relevant aglycons have two or even three hydroxyl groups. Depending on the site of the glycosidic linkage, the resulting conjugates may show considerable differences in their reactivity. The importance of glycoconjugation for the formation of aroma volatiles is demonstrated in the case of vitispirane **6** formation. Whereas the free aglycon **5A** was found to yield a whole pattern of volatile products, among which isomeric vitispiranes **6** were only present in minor quantities (15 %), the glucoconjugated form **5** almost exclusively forms the target compounds **6**. Glycosidation obviously stabilizes the hydroxyfunction at C-3 and, hence, cyclization to spiroether **6** is now the preferred reaction (*16*).

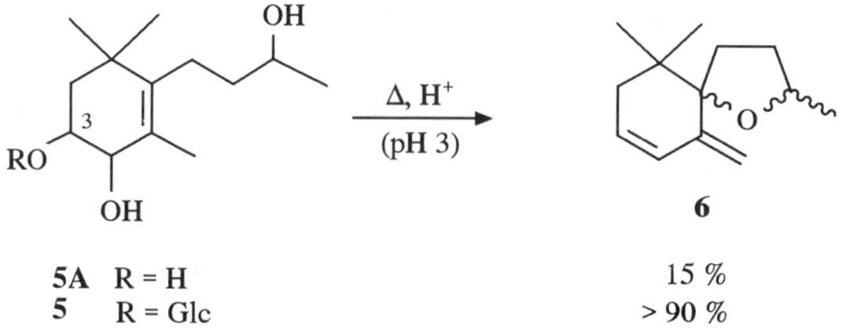

5A R = H	15 %
5 R = Glc	> 90 %

Figure 2. Influence of glycoconjugation on the rate of product formation, example vitispirane **6** formation from nonvolatile precursors **5** and **5A**.

Moreover, for the different classes of wine aroma constituents, i.e. for mono-
and norterpenoids as well as shikimic acid derivatives, multiple conjugating
moieties (ß-D-glucopyranosides, 6-O-α-L-rhamnopyranosyl-ß-D-glucopyranosides,
6-O-α-L-arabinofuranosyl-ß-D-glucopyranosides, and 6-O-ß-D-apiofuranosyl-ß-D-
glucopyranosides) have been determined in wine (*11,13*). As a further glycon
moiety, α-D-glucose has recently been identified. The newly identified phenyl-
ethyl-α-D-glucopyranoside **7** was present as a minor constituent in the glycosidic
fraction of Riesling wine (*48*). Due to the specificity of the cleaving enzymes,
precise information about the glycon part is required.

Differences observed in the composition of acid and enzymatic hydrolysates of
wines have led to speculations about the presence of glycoconjugates that may be in
part or fully resistant to enzymatic cleavage reactions (*11,28*). One example is
2-ethyl-3-methylmaleimide. This apparently chlorophyll-derived aroma compound,
which has been identified as an aglycon in Chardonnay grapes, was mainly liberated
by acid hydrolysis (*22*). From Riesling wine, we could recently isolate the known
N-glucoside **8** as its likely genuine precursor (*49,50*). This finding indicates that in
addition to the common O-glycosides other aroma precursors which may not be
susceptible to enzymatic cleavage reactions have to be expected to occur in wine.

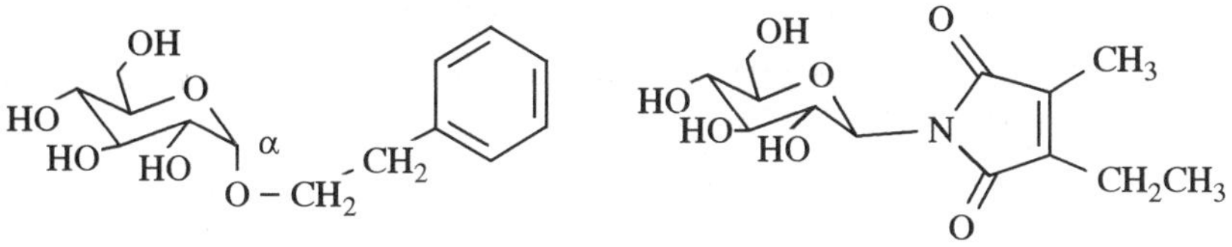

Figure 3. Structures of two newly isolated glycoconjugates from Riesling wine.

To avoid the above mentioned problems which are due to side activities of
commercial glycosidase preparations and the specificity of glycosidases for both, the
glycon as well as the aglycon moiety, an isolation and structural determination of
individual constituents in a precursor fraction should be attempted. This requires the
availability of preparative separation techniques that enable a gentle isolation of
reactive aroma precursors from the complex glycosidic fraction of wine.

Application of Countercurrent Chromatography to the Analysis of Reactive Aroma Precursors in Wine.

In recent years, significant improvements have been made to enhance the
performance and the efficiency of countercurrent chromatography (CCC). Besides
the previously used 'hydrostatic' techniques of *Rotation Locular Countercurrent
Chromatography* (RLCCC) and *Droplet Countercurrent Chromatography* (DCCC)
more recently developed, highly efficient 'hydrodynamic' techniques such as, e.g.,
Multi-Layer Coil Countercurrent Chromatography (MLCCC), are now available for
the separation and purification of complex natural mixtures. Especially for labile
natural compounds, such as, e.g., aroma precursors, CCC offers additional or
alternative procedures to the more extensively employed chromatographic
separations on solid stationary phases. Major advantages of CCC that have to be
stressed are:
(i) the absence of solid adsorbents, i.e. adsorption losses and the formation of
artifacts caused by active surfaces are eliminated.
(ii) Instead of solid packing materials, which in many cases are very costly, CCC
techniques rely exclusively on inexpensive solvent mixtures.

(iii) Large sample quantities (several grams per separation) can be applied and (iv) a total recovery of the sample material is guaranteed.

For a successful separation, all that is required is basically an immiscible solvent pair in which the components of the mixture have different partition coefficients according to the Nernst distribution law. Details about the instrumentation as well as numerous applications, including the separation of aroma precursors, can be found in the literature cited (23-28).

Due to its separation power, the technique of multi-layer coil countercurrent chromatography (MLCCC) has been used for the purification of a glycosidic XAD-2 isolate (20 g) which has been obtained from 100 L of a dealcoholized German Riesling wine. The initial preparative fractionation of the isolate was achieved on a '*preparative coil*' (75 m x 2.6 mm i.d. PTFE tubing) employing $CHCl_3/MeOH/H_2O$ (7:13:8) as solvent mixture. The separation was checked by TLC and fractions with similar R_f-values were pooled in seven combined fractions. These subfractions were then further purified with the '*analytical coil*' (160 m x 1.6 mm i.d. PTFE tubing) using EtOAc/n-BuOH/H_2O (3:2:5) as solvent system (27). After acetylation (Ac_2O/pyridine) and flash chromatography, the Riesling glycosides were finally purified by normal phase HPLC.

Identification of Novel Aroma Precursors in Riesling Wine

Isolation of the Glucose ester of (*E*)-2,6-dimethyl-6-hydroxy-2,7-dienoic acid. Of the many glycoconjugates isolated, one in particular showed an unusually low chemical shift for the anomeric proton. Whereas in ß-D-glucosides the anomeric proton resonates around δ 4.5 ppm, the anomeric proton in structure **9** showed a downfield shift and resonated at δ 5.7 ppm. This δ-value is typical for glucose esters (29,30). Additional signals in the ^{1}H-NMR spectrum of **9** included four olefinic protons, i.e. a typical ABX pattern for a vinyl group at δ 5.10, 5.23 and 5.90 ppm (J_{AB} = 1.2 Hz; J_{AX} = 10.5 Hz, *cis*-coupling; J_{BX} = 17.5 Hz, *trans*-coupling) as well as a methine proton at δ 6.86 ppm. The latter showed in addition to the coupling to H_2-4 (*J* = 7.0 Hz) a long-range coupling (*J* = 1.5 Hz) to the allylic methyl group at C2. The methylene groups at C4 and C5 resonated as multiplets at δ 2.23 and 1.65 ppm, respectively. Two three-proton singlets at δ 1.31 and 1.81 ppm were assigned to a tertiary methyl group attached to a carbon bearing a hydroxyl group (C6) and an allylic methyl group (Me-2), respectively. The ^{1}H NMR data for the terpene moiety are in good agreement with those published for 2,6-dimethyl-6-hydroxyocta-2,7-dienoic acid **9A** isolated from *Artemisia sieberi* (31). Additional spectral data for the novel wine constituent **9** have been published elsewhere (32).

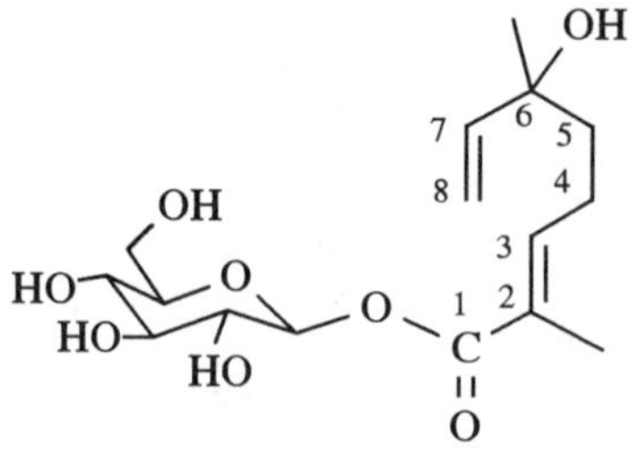

9

Figure 4. Structure of the newly identified glucose ester **9** from Riesling wine.

Whereas the glucose ester **9** has been identified for the first time as a natural wine constituent, glycoconjugates of its reduced form, i.e. of the monoterpene diol **11**, are known Riesling wine constituents (*2*). Under acidic conditions, diol **11** was partially converted into the bicyclic ether **12**, the so-called dillether (*2*). In analogy to the formation of ether **12** from terpene diol **11**, a likely formation of lactone **10** from acid **9A** could be be expected (cf. Fig. 5). This so-called wine-lactone **10**, first identified as an essential oil metabolite in the Koala (*33*), has recently been established by Guth (*34*) as a major aroma contributor in two white wine varieties. The 3*S*,3a*S*,7a*R*-configured isomer of **10**, which has been identified in wine, is reported to possess an unusual low flavor threshold of 0.01-0.04 pg/L of air and a *'sweet, coconut-like'* aroma (*35*).

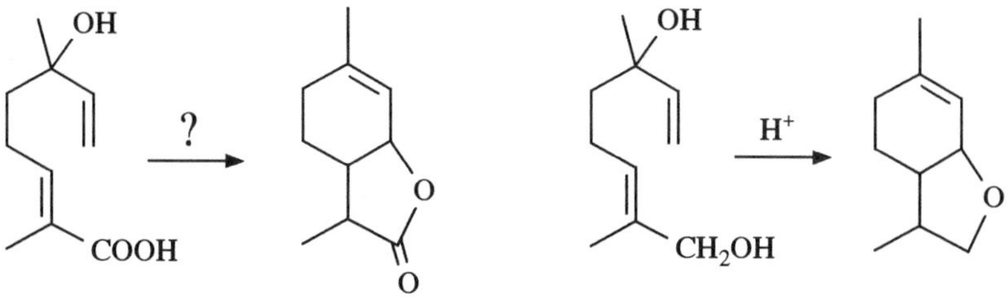

Figure 5. Postulated formation of wine-lactone **10** from monoterpenoid acid **9A** in analogy to dillether **12** formation from the structurally related diol **11**.

In order to substantiate the hypothetic pathway for wine-lactone **10** formation, the presumed precursor **9A** has been synthesized (cf. Fig. 6). SeO$_2$ oxidation of linalyl acetate **13** yielded aldehyde **14** which was converted into the carboxylic methyl ester **15** by a cyanide-catalyzed oxidative esterification (*36*). Deprotection of **15** was achieved under mild conditions using porcine liver esterase (PLE). It is noteworthy that after PLE mediated hydrolysis, trace amounts of wine-lactone **10** could be identified in the reaction mixture. After purification of acid **9A**, aliquots have been subjected to thermal treatment at pH 3.2, 2.5 and 2.0, respectively. In all cases, wine-lactone **10** was detectable as conversion product of acid **9A**. The structure elucidation of additional degradation products (MS spectral data are gathered in Tab. I) as well as long term storage experiments (i.e. degradation of **9A** in model wine medium at 40°C) are subjects of ongoing studies.

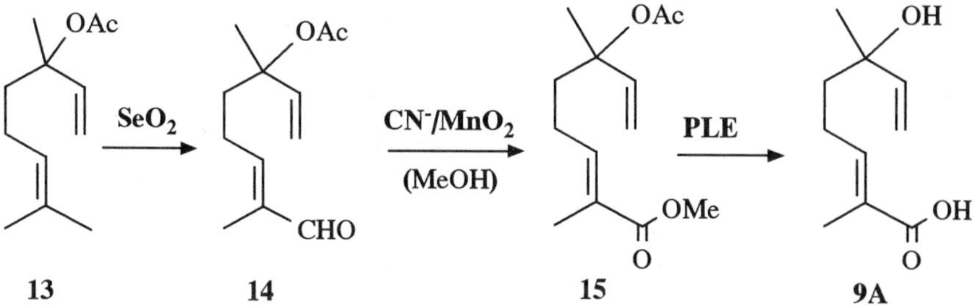

Figure 6. Synthesis of (*E*)-2,6-dimethyl-6-hydroxyocta-2,7-dienoic acid **9A** from linalyl acetate **13** (for details cf. text).

Table I. Mass Spectral Data (70 eV) of Major Degradation Products of Acid 9A

	R_i (DB-5)*	m/z (%)
Unknown A	1385	166 (1), 148 (6), 137 (2), 133 (3), 121 (34), 111 (7), 105 (25), 93 (57), 91 (33), 79 (37), 67 (40), 53 (39), 41 (100).
Unknown B		
(*1st isomer*)	1431	166 (19), 148 (11), 133 (11), 121 (86), 111 (53), 105 (76), 98 (64), 93 (81), 91 (88), 79 (100), 65 (29), 53 (45), 41 (65).
(*2nd isomer*)	1447	166 (35), 151 (5), 148 (5), 133 (9), 121 (77), 111 (89), 105 (86), 98 (100), 93 (98), 91 (96), 79 (100), 65 (36), 53 (50), 41 (79).
Lactone **10**	1456	166 (19), 151 (100), 138 (9), 123 (14), 107 (32), 93 (72), 79 (44), 69 (14), 55 (34), 41 (24).
Unknown C		
(*1st isomer*)	1517	166 (29), 151 (6), 133 (5), 121 (100), 105 (60), 91 (41), 77 (28), 65 (11), 53 (14), 41 (22).
(*2nd isomer*)	1541	166 (31), 151 (6), 133 (6), 121 (100), 105 (61), 91 (36), 77 (28), 65 (11), 53 (16), 41 (21).

*Linear retention index on a J&W DB-5 capillary column (30 m x 0.25 mm i.d., film thickness 0.25 μm).

Isolation of Two Glucosidic Precursors of ß-Damascenone From Riesling Wine. Another important aroma compound of Riesling wine is the norisoprenoid ketone ß-damascenone **19** with an aroma threshold of 2 pg/g in water (*37*). By using MLCCC as well as HPLC, two glucoconjugates of 3-hydroxy-7,8-didehydro-ß-ionol could be isolated and purified from Riesling wine (cf. Fig. 7). The site of the glycosidic linkage was in each case established from heteronuclear multi-bond correlation (HMBC) NMR experiments. The complete set of spectroscopic data has been published elsewhere (Baderschneider, B.; Skouroumounis, G.; Winterhalter, P. *Nat. Prod. Lett.*, in press).

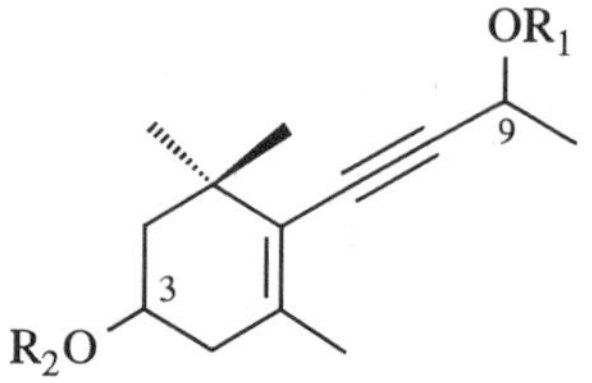

	R_1	R_2
16	Glc	H
17	H	Glc
18	H	H

Figure 7. Structures of two acetylenic precursors of ß-damascenone **19** isolated from Riesling wine.

In acidic medium, the acetylenic diol **18** as well as its glucoconjugated form **16** have been demonstrated to undergo dehydration as well as a Meyer-Schuster rearrangement, which generates ß-damascenone **19** and 3-hydroxy-ß-damascone **20** (*38,39*). Contrary to ketone **19** which is a key flavor compound in many natural products, the hydroxy-derivative **20** is known to be odorless. Thus, for the aroma of wines, maximum concentrations of ketone **19** are desirable. In this regard, it has to be stressed that the site of glycosidation significantly influences the reactivity of the aroma conjugates as well as the relative proportions of volatiles formed. For the 9-O-glucoconjugate **16**, kinetic studies of Skouroumounis *et al.* (*39*) have shown that - compared to the free aglycon **18** - a greater proportion of the target ketone **19** is formed (cf. Fig. 8). For the 9-O-glucoside **16**, it is assumed that through stabilization at C-9, dehydration at C-3 is favored, thus explaining the observed higher yields of ß-damascenone **19**. *Vice versa*, for the 3-O-glucoside **17** - through stabilization of the hydroxyl-function at C-3 - it is expected a higher amount of hydroxyketone **20** will be obtained. Compound **20** was found to be stable under pH conditions of wine, neither the free aglycon nor its glucoside will undergo further transformations to give ß-damascenone **19**. Consequently, of the two newly identified glucosides, the 9-O-conjugate **16** has to be regarded as the more important progenitor of ß-damascenone **16** in Riesling wine.

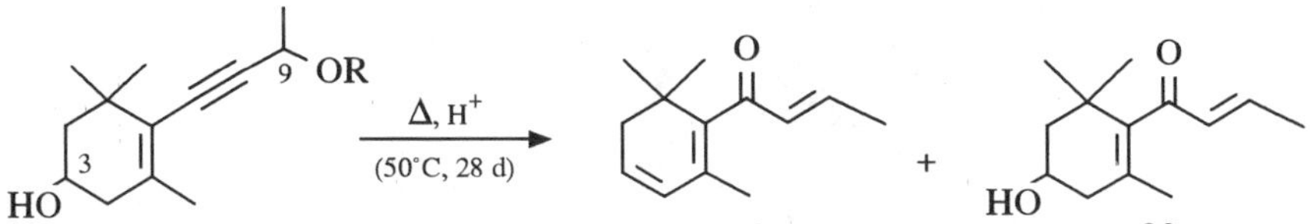

16 (R = Glc)	10 %	90 %
18 (R = H)	5 %	95 %

Figure 8. Influence of glycoconjugation on the rate of reaction products **19** and **20** formed from acetylenic diol **18** and its 9-O-glucoconjugate **16** according to Skouroumounis *et al.* (*39*).

Isolation of Additional Glycosides from Riesling Wine.

In addition to the aforementioned aroma precursors, further glycoconjugates have been isolated and characterized from Riesling wine during this study. Completely characterized glycosides with mono- and norterpenoid, benzylic and aliphatic aglycon moieties are depicted in Fig. 9. Spectral data for the newly identified aliphatic glucosides **21** and **22** as well as the norisoprenoid conjugate **32** are gathered in Table II. Spectral data for the known wine constituents **23-31** can be found in the literature cited (*40-45*).

Table II. Spectral Data for Riesling Glucoconjugates 21, 22 and 32.

21	**DCI-MS** (reactant gas: NH$_3$) pseudo-molecular ion at m/z 436 [M(418)+NH$_4$]$^+$; **^{1}H-NMR** (360 MHz, CDCl$_3$, ppm, J in Hz): δ 0.87 and 0.88 (2 x 3H, 2d, J = 6.6, 2CH$_3$-C3); 1.27-1.54 (2H, m, H$_2$C2); 1.65 (1H, m, HC3); 1.99, 2.01, 2.02, 2.08 (4 x 3H, 4s; acetates); 3.50 (1H, dt, J = 6.9, 9.7, H$_a$C1); 3.68 (1H, ddd, J = 9.9, 4.7, 2.5, HC5'); 3.89 (1H, dt, J = 6.3, 9.7, H$_b$C1); 4.12 (1H, dd, J = 12.3, 2.4, H$_a$C6'); 4.25 (1H, dd, J = 12.3, 4.7, H$_b$C6'); 4.47 (1H, d, J = 8.0, HC1'); 4.97 (1H, dd, J = 9.6, 8.0, HC2'); 5.07 (1H, dd, J = 9.7, 9.7, HC4'); 5.18 (1H, dd, J = 9.5, 9.5, HC3'). ^{13}C-NMR (63 MHz, CDCl$_3$, ppm): δ 20.5 - 20.6 (acetates), 22.28 and 22.54 (2Me-C3), 24.85 (C3), 38.19 (C2), 62.18 (C6'), 68.53 (C4'), 68.78 (C1), 71.55 (C2'), 71.88 (C5'), 73.03 (C3'), 100.90 (C1'), 169.1 - 170.5 (acetates).
22	**DCI-MS** (reactant gas: NH$_3$) pseudo-molecular ion at m/z 436 [M(418)+NH$_4$]$^+$; **^{1}H-NMR** (360 MHz, CDCl$_3$, ppm, J in Hz): δ 0.85- 0.89 (6H, m, CH$_3$-C2 and CH$_3$-C3); 1.13 (1H, ddq, J = 7.4, 7.4, 13.8, H$_a$C3); 1.52 (1H, ddq, J = 6.9, 6.9, 13.8, H$_b$C3); 1.64 (1H, m, HC2); 2.01, 2.02, 2.03, 2.09 (4 x 3H, 4s; acetates); 3.20 (1H, dd, J = 7.2, 9.4, H$_a$C1); 3.68

Table II (cont.)

(1H, ddd, J = 10.1, 4.7, 2.5, HC5'); 3.79 (1H, dd, J = 5.5, 9.4, H_bC1); 4.13 (1H, dd, J = 12.2, 2.4, H_aC6'); 4.26 (1H, dd, J = 12.2, 4.7, H_bC6'); 4.49 (1H, d, J = 8.0, HC1'); 5.00 (1H, dd, J = 9.5, 8.0, HC2'); 5.09 (1H, dd, J = 9.5, 9.5, HC4'); 5.20 (1H, dd, J = 9.5, 9.5, HC3'). ^{13}C-NMR (63 MHz, CDCl$_3$, ppm): δ 20.5 - 20.6 (acetates), 11.18 (C4), 16.41 (C5), 25.92 (C3), 34.89 (C2), 62.18 (C6'), 68.81 (C4'), 71.88 (C2'), 73.00 (C3'), 73.19 (C5'), 75.21 (C1), 101.28 (C1'), 169.1 - 170.6 (acetates).

32   **DCI-MS** (reactant gas: NH$_3$) pseudo-molecular ion at m/z 574 [M(556)+NH$_4$]$^+$; ^{1}**H-NMR** (360 MHz, CDCl$_3$, ppm, J in Hz): δ 1.05 and 1.08 (2 x 3H, 2s, 2CH$_3$-C1); 1.16 (3H, d, J = 6.4, CH$_3$-C9); 1.4-2.2 (4H, m, H$_2$C7/H$_2$C8); 1.99 (3H, d, J = 1.3, CH$_3$-C5); 2.01, 2.03, 2.04, 2.09 (4 x 3H, 4s; acetates); 2.22 (1H, d, J = 17, H$_a$C2); 2.42 (1H, d, J = 17, H$_b$C2); 3.63 (1H, m, HC5'); 3.65 (1H, m, HC9); 4.18 (1H, dd, J = 12.3, 2.4, H$_a$C6'); 4.22 (1H, dd, J = 12.3, 4.7, H$_b$C6'); 4.49 (1H, d, J = 8.0, HC1'); 4.92 (1H, dd, J = 9.5, 8.0, HC2'); 5.05 (1H, dd, J = 9.5, 9.5, HC4'); 5.10 (1H, dd, J = 9.5, 9.5, HC3'); 5.83 (1H, brs, HC4). ^{13}C-NMR (63 MHz, CDCl$_3$, ppm): δ 18.78 (Me-C5), 20.2 - 20.9 (acetates), 20.90 (Me-C9), 22.90 and 24.14 (2Me-C1), 32.04 (C8), 34.04 (C7), 41.17 (C1), 50.12 (C2), 61.90 (C6'), 68.65 (C4'), 71.60 (C2'), 72.05 (C5'), 72.88 (C3'), 76.36 (C9), 78.99 (C6), 99.88 (C1'), 126.30 (C4), 162.50 (C5), 169.3 - 170.7 (acetates), 197.50 (C3).

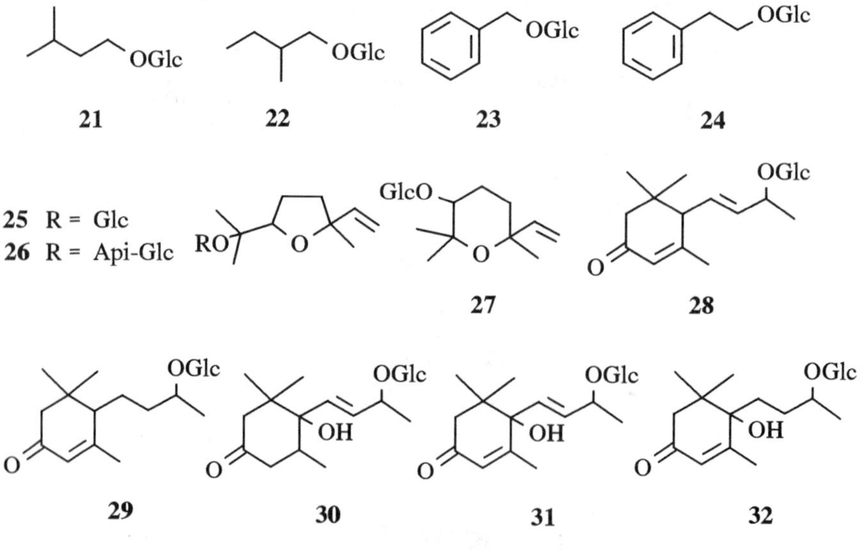

Figure 9. Structures of additional glycoconjugates isolated from Riesling wine during this study: ß-D-glucopyranosides of 3-methylbutanol **21**, 2-methyl-butanol **22**, benzyl alcohol **23**, 2-phenylethanol **24**, furanoid linalool oxides (two diastereoisomers) **25**, pyranoid linalool oxides (two diastereoisomers) **27**, 3-oxo-7,8-dihydro-α-ionol **28**, 3-oxo-α-ionol **29**, 4,5-dihydro-vomifoliol **30**, vomifoliol **31**, and 7,8-dihydro-vomifoliol as well as the 6-O-ß-D-apiofurano-syl-ß-D-glucopyranosides of furanoid linalool oxides (two diastereoisomers).

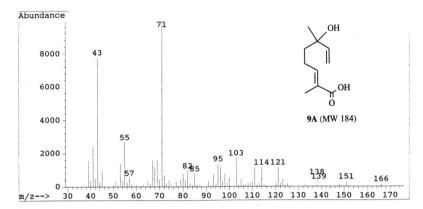

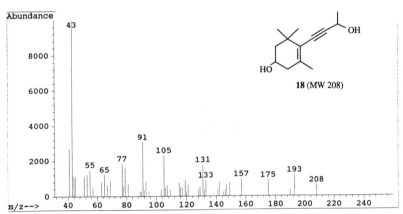

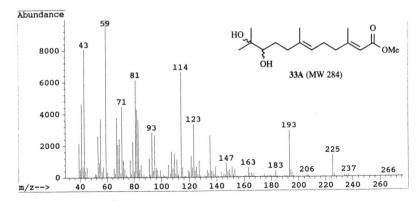

Figure 10. Mass spectral data (70eV) of Riesling aglycons **9A** and **18**, as well as methylated **33A**.

Isolation of the Glucose Ester of 10,11-Dihydroxy-3,7,11-trimethyl-2,6-dodeca-dienoic Acid from Riesling wine. During our studies on aroma precursors in Riesling wine, we have also isolated other secondary metabolites which obviously are not involved in flavor formation. An interesting example is the farnesene derivative **33**. This structure with a fifteen carbon skeleton has been isolated as glucose ester **33** from the glycosidic XAD-2 isolate. It has been completely characterized using one and two dimensional NMR techniques (Winterhalter, P.; Baderschneider, B.; Bonnländer, B. submitted to *J. Agric. Food Chem.*). The structure of the methylated aglycon was furthermore confirmed by converting the commercially available juvenile hormone III into diol **33A** (cf. Fig. 11). Whereas the specific role of glucose ester **33** remains to be elucidated, one can speculate about its possible implication in the formation of other grape and wine constituents. Farnesene derivatives have been discussed as a possible biogenetic source of abscisic acid (ABA) (*46,47*). The latter has also been isolated and characterized from Riesling wine in the present study.

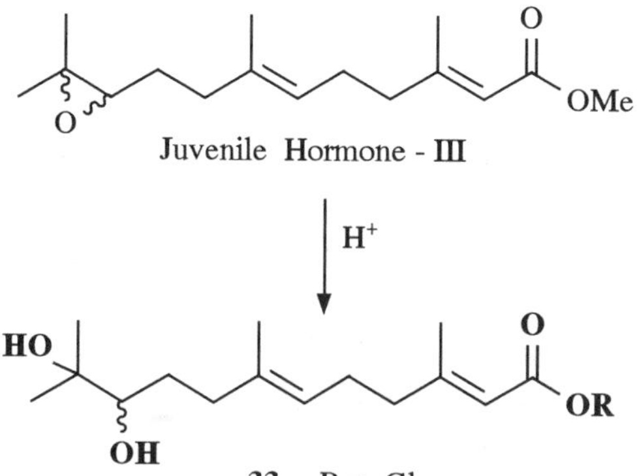

Juvenile Hormone - III

Figure 11. Structure of the novel glucose ester **33** and the syntheses of the aglycon **33A** (methyl ester) through acid catalyzed conversion of juvenile hormone III.

Conclusions

Due to the gentle isolation conditions, the application of CCC techniques in natural product analysis is steadily increasing. It has been demonstrated that MLCCC facilitates the isolation of aroma-relevant glycoconjugates from the complex glycosidic mixture of Riesling wine. The intact glycoconjugates are required to study their specific role in wine flavor formation. However, CCC is not restricted to these studies on aroma precursors, it is equally important for elucidating the structure of other polar wine constituents, such as, e.g., polyphenols. Research in the area of antioxidative constituents in Riesling wine is presently under active investigation.

Acknowledgments

The skillful assistance of M. Messerer is gratefully acknowledged. Dr. G. Skouroumounis is thanked for his helpful comments on the ß-damascenone studies. The Deutsche Forschungsgemeinschaft, Bonn, is thanked for funding the research.

Literature Cited

1. Strauss, C.R.; Gooley, P.R.; Wilson, B.; Williams, P.J. *J. Agric. Food Chem.* **1987**, *35*, 519-514.
2. Strauss, C.R.; Wilson, B.; Williams, P.J. *J. Agric. Food Chem.* **1988**, *36*, 569-573.
3. Winterhalter, P., Sefton, M.A.; Williams, P.J. *J. Agric. Food Chem.* **1990**, *38*, 1041-1048.
4. Winterhalter, P., Sefton, M.A.; Williams, P.J. *Am. J. Enol. Vitic.* **1990**, *41*, 277-283.
5. Winterhalter, P. *J. Agric. Food Chem.* **1991**, *39*, 1825-1829.
6. Waldmann, D.; Winterhalter, P. *Vitis* **1992**, *31*, 169-174.
7. Full, G.; Winterhalter, P. *Vitis* **1994**, *33*, 241-244.
8. Marinos, V.A.; Tate, M.E.; Williams, P.J. **1994**, *42*, 2486-2492.
9. Williams, P.J.; Sefton, M.A.; Francis, I.L. In *Flavor Precursors - Thermal and Enzymatic Conversions*; Teranishi, R.; Takeoka, G.R.; Güntert, M., Eds.; ACS Symp. Ser. 490; American Chemical Society: Washington, DC, 1992, pp. 74-86.
10. Winterhalter, P. In *Flavor Precursors - Thermal and Enzymatic Conversions*; Teranishi, R.; Takeoka, G.R.; Güntert, M., Eds.; ACS Symp. Ser. 490; American Chemical Society: Washington, DC, 1992, pp. 98-115.
11. Williams, P.J. In *Flavor Science - Sensible Principles and Techniques*; Acree, T.E.; Teranishi, R., Eds.; American Chemical Society: Washington, DC, 1993, pp. 287-303.
12. Williams, P.J.; Sefton, M.A.; Marinos, V.A. In *Recent Developments in Flavor and Fragrance Chemistry*; Hopp, R.; Mori, K., Eds.; VCH Verlagsgesellschaft: Weinheim, Germany, 1993, pp. 283-290.
13. Günata, Z.; Dugelay, I.; Sapis, J.C.; Baumes, R.; Bayonove, C. In *Progress in Flavour Precursor Studies - Analysis, Generation, Biotechnology*; Schreier, P.; Winterhalter, P., Eds.; Allured Publ.: Carol Stream, IL, 1993, pp. 219-234.
14. Winterhalter, P. In *Progress in Flavour Precursor Studies - Analysis, Generation, Biotechnology*; Schreier, P.; Winterhalter, P., Eds.; Allured Publ.: Carol Stream, IL, 1993, pp. 31-44.
15. Stahl-Biskup, E.; Intert, F.; Holthuijzen, J.; Stengele, M.; Schulz, G. *Flav. Fragr. J.* **1993**, *8*, 61-80.
16. Winterhalter, P.; Skouroumounis, G.K. In *Advances in Biochemical Engineering/Biotechnology - Vol. 55: Biotechnology of Aromas*; Berger, R.G., Ed.; Springer Verlag: Heidelberg, Germany, 1997, pp. 73-105.
17. Rapp, A.; Güntert, M.; Ullemeyer, H. *Z. Lebensm. Unters. Forsch.* **1985**, *180*, 109-116.
18. Winterhalter, P. In *Connaissance Aromatique des Cépages et Qualité des Vins*; Bayonove, C.; Crouzet, J.; Flanzy, C.; Martin, J.C.; Sapis, J.C., Eds.; Revue Française d'Œnologie: Lattes, France, 1994, pp. 65-73.
19. Williams, P.J.; Strauss, C.R.; Wilson, B.; Massy-Westropp, R.A. *J. Chromatogr.* **1982**, *235*, 471-480.
20. Günata, Y.Z.; Bayonove, C.L.; Baumes, R.L.; Cordonnier, R.E. *J. Chromatogr.* **1985**, *331*, 83-90.
21. Sefton, M.A.; Williams, P.J. *J. Agric. Food Chem.* **1991**, *39*, 1994-1997.
22. Sefton, M.A.; Francis, I.L.; Williams, P.J. *Am. J. Enol. Vitic.* **1993**, *44*, 359-370.
23. Ito, Y. *CRC Crit. Rev. Anal. Chem.* **1986**, *17*, 65-143.

24. Mandava, N.B.; Ito, Y. *Countercurrent Chromatography: Theory and Practice*; Marcel Dekker: New York, 1988.
25. Conway, W.D. *Countercurrent Chromatography - Apparatus, Theory & Applications*; VCH: Weinheim, Germany, 1990.
26. *Modern Countercurrent Chromatography*; Conway, W.D.; Petroski, R.J., Eds.; ACS Symp. Ser. 593; American Chemical Society: Washington, DC, 1995.
27. Roscher, R.; Winterhalter, P. *J. Agric. Food Chem.* **1993**, *41*, 1452-1457.
28. Krammer, G.E.; Buttery, R.G.; Takeoka, G.R. In *Fruit Flavors - Biogenesis, Characterization and Authentication*; Rouseff, R.L.; Leahy, M.M., Eds; ACS Symp. Ser. 596; American Chemical Society: Washington, DC, 1995, pp. 164-181.
29. Koshimizu, K.; Inui, M.; Fukui, H.; Mitsui, T. *Agric. Biol. Chem.* **1986**, *32*, 789-791.
30. Winterhalter, P.; Lutz, A.; Schreier, P. *Tetrahedron Lett.* **1991**, *32*, 3669-3670.
31. Marco, J.A.; Sanz-Cervera, J.F.; Sancenon, F.; Jakupovic, J.; Rustaiyan, A.; Mohamadi, F. *Phytochemistry* **1993**, *34*, 1061-1065.
32. Winterhalter, P.; Messerer, M.; Bonnländer, B. *Vitis* **1997**, *36*, 55-56.
33. Southwell, I.A. *Tetrahedron Lett.* **1975**, *16*, 1885-1888.
34. Guth, H. *Lebensmittelchemie* **1995**, *49*, 107.
35. Guth, H. *Helv. Chim. Acta* **1996**, *79*, 1559-1571.
36. Corey, E.J.; Gilman, N.W.; Ganem, B.E. *J. Am. Chem. Soc.* **1968**, *90*, 5616-5617.
37. Buttery, R.G.; Teranishi, R.; Ling, L.C. *Chem. & Ind. (London)* **1988**, 238.
38. Skouroumounis, G.K.; Massy-Westropp, R.A.; Sefton, M.A.; Williams, P.J. *J. Agric. Food Chem.* **1995**, *43*, 974-980.
39. Skouroumounis, G.K.; Massy-Westropp, R.A.; Sefton, M.A.; Williams, P.J. In *Progress in Flavour Precursor Studies - Analysis, Generation, Biotechnology*; Schreier, P.; Winterhalter, P., Eds.; Allured Publ.: Carol Stream, IL, 1993, pp. 275-279.
40. Williams, P.J.; Strauss, C.R.; Wilson, B.; Massy-Westropp, R.A. *Phytochemistry* **1983**, *22*, 2039-2041.
41. Strauss, C.R.; Wilson, B.; Williams, P.J. *Phytochemistry* **1987**, *26*, 1995-1997.
42. Sefton, M.A.; Winterhalter, P.; Williams, P.J. *Phytochemistry* **1992**, *31*, 1813-1815.
43. Baumes, R.; Dugelay, I.; Günata, Y.Z.; Tapiero, C.; Bitteur, S.; Bayonove, C. In *Connaissance Aromatique des Cépages et Qualité des Vins*; Bayonove, C.; Crouzet, J.; Flanzy, C.; Martin, J.C.; Sapis, J.C., Eds.; Revue Française d'Œnologie: Lattes, France, 1994, pp. 90-98.
44. Baumes, R.; Aubert, C.C.; Günata, Y.Z.; De Moor, W.; Bayonove, C.L.; Tapiero, C. *J. Essent. Oil Res.* **1994**, *6*, 587-599.
45. Baltenweck-Guyot, R.; Trendel, J.M.; Albrecht, P.; Schaeffer, A. *Phytochemistry* **1996**, *43*, 621-624.
46. Neill, S.J.; Horgan, R. *Phytochemistry* **1983**, *22*, 2469-2472.
47. Bennett, R.D.; Norman, S.M.; Maier, V.P. *Phytochemistry* **1990**, *29*, 3473-3477.
48. Watanabe, N.; Messerer, M.; Winterhalter, P., *Nat. Prod. Lett.* **1997**, *10*, 39-42.
49. Krajewski, D.; Tóth, G.; Schreier, P. *Phytochemistry* **1996**, *43*, 141-143.
50. Baderschneider, B.; Winterhalter, P. *Vitis* **1997**, *36*, 159-160.

Chapter 2

The Contribution of Glycoside Precursors to Cabernet Sauvignon and Merlot Aroma

Sensory and Compositional Studies

I. Leigh Francis[1], Stella Kassara[1], Ann C. Noble[2], and Patrick J. Williams[1]

[1]The Australian Wine Research Institute, P.O. Box 197, and Cooperative Research Centre for Viticulture, P.O. Box 145, Glen Osmond, South Australia 5064, Australia
[2]Department of Viticulture and Enology, University of California, Davis, CA 95616

Volatile compounds released from Cabernet Sauvignon and Merlot grape glycoside fractions, isolated from both skin and juice, were studied by sensory descriptive analysis and by gas chromatography-mass spectrometry (GC-MS). Both acid- and enzyme-hydrolysates were studied. The contribution to wine aroma of the different fractions was evaluated by sensory analysis of white wines to which the hydrolysates had been added. Acid-hydrolysates from each variety increased the intensity of attributes such as *tobacco*, *chocolate* and *dried fig*. In contrast, glycosidase enzyme-hydrolysates gave no detectable change in aroma. The relationship among the aroma attributes of the hydrolysates and their volatile composition was investigated using partial least square regression analysis (PLS), which indicated that the intensity of the attributes *dried fig*, *tobacco* and *honey* could be related to the concentration of specific compounds of the norisoprenoid, benzene derivative, monoterpene and aliphatic classes. The red-free glycosyl-glucose (G-G) concentration of the skin extracts and juices was correlated with the scores of aroma attributes of the glycoside hydrolysates, suggesting the potential of the G-G assay as a predictor of wine aroma.

The awareness that glycosidically-conjugated volatile compounds are present in grape berries and other fruits has stimulated substantial research interest in these constituents. Research on glycosidic flavor precursors has been the subject of several reviews *(1-6)*. In the case of wine grapes, it is becoming evident from sensory studies that grape berry derived glycosidically-bound volatile compounds are capable of making a contribution to varietal wine flavor.

For the non-floral white grape varieties Chardonnay, Semillon and Sauvignon

Blanc, a connection has been established by sensory descriptive analyses between the aroma attributes of hydrolyzed flavor precursors from the grapes and wines of these varieties *(7-10)*. These studies have demonstrated that grape glycosides are of importance to white wine flavor, in particular after a period of wine storage. Similarly, for the black grape variety Shiraz, a sensory study has indicated that juice glycosidic hydrolysates have aroma characteristics in common with those of wines of that variety *(11)*.

Numerous volatiles are released upon hydrolysis of glycoside isolates *(1, 13)*, many of which are presumed to be acting as flavor compounds. Different grape varieties apparently produce glycosides which, when hydrolyzed, release differing proportions of monoterpenes, C_{13} norisoprenoids and benzene derivatives, as well as other volatiles. However, there is little reliable aroma threshold information regarding many of these compounds, and there has been no systematic attempt to relate the volatile composition of the hydrolysates to their sensory properties.

The present work was undertaken to explore the contribution that glycosylated volatiles of black grapes can make to red wine aroma, and to attempt to identify those compounds or classes of compounds which may be responsible for specific aroma attributes. Cabernet Sauvignon and Merlot were the varieties chosen for the study.

Materials and Methods

Grapes and wines. Grapes from the 1994 vintage were picked at commercial ripeness from vineyards in both California and South Australia. The fruit taken for these experiments and their composition are listed in Table I.

Table I. Juice composition of grape samples used for isolation of glycosides.

Variety	Source	°Brix	pH	Titratable acidity[a](g/L)
Cabernet Sauvignon	Coonawarra (South Australia)	23.5	3.48	5.1
	Davis (California)	24.1	3.48	4.8
	Napa Valley (California)	23.0	3.26	5.4
Merlot	Lenswood (South Australia)	23.8	3.33	5.0
	Davis (California)	25.2	3.94	4.3
	Napa Valley (California)	23.4	3.31	5.1

[a]As tartaric acid

The fruit was crushed and destemmed, followed by a light pressing in a basket press (Californian fruit) or a water bag press (Australian fruit). The skins were stored separately from the expressed juice, with all material held frozen at less than -10°C.

Wines, which were made from separate lots of the Napa Merlot and Cabernet Sauvignon fruit, were vinified at UC Davis, remaining on skins until approximately 5° Brix. The base wine used in the sensory study was a 1993 Napa Chardonnay also made at UC Davis. None of the wines had oak treatment or went through malolactic fermentation. The wines were bottled into 750 mL clear glass bottles sealed with screw cap closures.

Sample preparation. To estimate glycoside extraction during winemaking, the grape skins (2.1 kg lots) were subjected to an extraction procedure involving contact with model wine solution (prepared as described in *(7)*, 2.4 L) for 7 days at 23-25°C, with periodic agitation. Both the skin extracts and juices were centrifuged and the supernatant filtered through a 5 μm membrane.

Isolation and preparation of glycosides for hydrolysis, including solvent extraction with Freon 11 to remove any free volatile compounds before hydrolysis, was performed as described previously *(7)*. Acid hydrolysis was performed on a glycosidic isolate in a volume of model wine, 1/25th that of the original sample volume. The solutions to be used for sensory analyses were transferred to glass teflon sealed screw cap bottles, while solutions for GC/MS analysis were transferred to glass ampoules, and heated at 50°C under a nitrogen atmosphere for 28 days. After this period, the solutions were cooled, and stored at -20°C until required for analysis.

For enzyme hydrolysis a glycosidic extract prepared from 1500 mL juice or skin extract was hydrolyzed in pH 5 buffer (162 mL) at 37°C for 16 h with Rohapect C (12 mg, Röhm, Darmstadt, Germany).

Glycosyl-Glucose (G-G) analysis. The skin extracts and juices were assayed for total G-G (3 mL of skin extracts, and 10 mL of juices taken for analysis) and anthocyanin concentration (1 mL taken for analysis) using procedures set out in Iland et al *(17)*.

Sensory analysis. Sensory descriptive analysis on the aroma of the 15 samples (see Table II) was undertaken as described previously *(7)* using 14 judges from the Department of Viticulture and Enology, UC Davis. The glycoside hydrolysate concentrates were diluted in base wine (BW) at double strength (ie twice the concentration of glycosides present in the original juice or skin extract sample) for sensory analysis. All assessments were done in May 1995 in duplicate and made in isolated booths under red light using black glasses to mask any color differences. The attributes that were rated by the panel were each defined by reference standards made up in base wine (Table III).

Compositional analysis. Juice or skin glycoside hydrolysates (equivalent to 250 mL of juice or skin extracts) were spiked with a standard solution of 1-octanol and 2,5-dimethylphenol in ethanol (to a concentration for skin hydrolysates 0.12 mg/L, for juice hydrolysates 0.06 mg/L) and extracted with 1:1 ether:pentane (3 x 10 mL). The organic extracts were dried with magnesium sulfate and concentrated by fractional distillation through a Vigreux column packed with Fenske helices prior to analysis. GC/MS analyses were carried out in duplicate as described previously *(16)*.

Table II. Summary of the fifteen samples taken for sensory descriptive analysis, including twelve juice and skin extract glycoside hydrolysates, two red wines, and the base wine.

Sample code	Variety	Source	Glycosides from:
ACS[a]	Cabernet Sauvignon	Coonawarra, Australia	skin extract
ACJ[a]	"	"	juice
NCS[a]	"	Napa, California	skin extract
NCJ[a]	"	"	juice
DCS[a]	"	Davis, California	skin extract
DCJ[a]	"	"	juice
Napa Cabernet wine	"	Napa, California	—[b]
AMS[a]	Merlot	Lenswood, Australia	skin extract
AMJ[a]	"	"	juice
NMS[a]	"	Napa, California	skin extract
NMJ[a]	"	"	juice
DMS[a]	"	Davis, California	skin extract
DMJ[a]	"	"	juice
Napa Merlot wine	"	Napa, California	—[b]
BW[c]	Chardonnay	Napa, California	—[b]

[a]presented for descriptive analysis diluted in the base wine (BW). [b]Glycosides not isolated from wines. [c]Base wine.

Table III. Aroma reference standards used and their composition

Attribute	Composition[a]
Floral	20 mL of a stock solution of rose petals (10 g) steeped in 500 mL base wine for 24 h, filtered, and 2-phenyl ethanol added (10 µL)
Apple	1/4 fresh peeled, sliced apple
Honey	2 mL honey
Berry	1 frozen raspberry (crushed), 2 g strawberry jam, 5 g blackberry jam
Dried fig	1 dried fig, cut into 1 cm^2 pieces
Chocolate	0.5 g dark chocolate shavings, 0.5 g cocoa powder (Hersheys)
Tobacco	few flakes of cigarette tobacco (Camel), tea bag soaked for 1 min

[a]Made up in 100 mL base wine.

Statistical analyses. Three-way analyses of variance treating judges as a random effect were performed on each descriptive term using SAS Institute Inc. JMP 3.1 (Cary, North Carolina). Principal component analysis of the correlation matrix of the mean intensity ratings was performed with Varimax rotation. Over 200 GC peaks were quantified, thus, to reduce the number of volatile compounds, several steps were undertaken to prescreen the GC/MS data. Firstly, those which did not vary significantly among the 12 samples (by one-way analysis of variance) were eliminated from further analysis. Secondly, one compound from each pair of compounds with highly significant correlation coefficients (r>0.85) was excluded from further analysis. Finally, inspection of the data showed that some compounds were present at substantially higher concentration in the enzyme hydrolysates than in any of the acid hydrolysis samples. With the knowledge that these enzyme treated samples did not contribute any detectable aroma when added to a base wine (see below), these particular compounds were also eliminated from further analysis. Partial Least Squares (PLS) regression analysis was used to relate the sensory data to the instrumental data. PLS2 was performed with cross validation on the normalized sensory and compositional data for the 12 acid hydrolysate samples using the Unscrambler (Camo A/S, Trondheim, Norway).

Results and Discussion

Glycosides were obtained from juice and skin extracts from both Cabernet Sauvignon and Merlot fruit, sourced from Australian and Californian vineyards. The glycoside isolates were acid hydrolyzed at elevated temperature in a model wine medium. This hydrolysis was carried out to simulate conditions, although in an accelerated manner, that could occur as wine is stored and matured, ie volatiles will be slowly produced from their non-volatile precursors.

Acid hydrolysates were added to a low aroma intensity white wine (ie the base wine), and the aroma properties of these samples were assessed by sensory descriptive analysis. In addition, the glycoside isolates from the Australian vineyards were subjected to glycoside hydrolase enzyme treatment, and duo-trio difference tests were performed on these hydrolysates added to a base wine. The volatile composition of each of the hydrolysates was investigated by GC/MS, and relationships between the two sets of data were determined. Finally, the glycoside concentration of each of the juices and skin extracts was determined by the glycosyl-glucose assay.

Sensory analysis. Significant differences in intensity were found for all seven aroma terms by analysis of variance (data not shown). Because of a highly significant judge-by-wine interaction, the berry term was excluded from further data analysis.

Figure 1 shows the mean ratings for the Napa Cabernet Sauvignon samples (juice glycoside hydrolysate, skin glycoside hydrolysate, and the wine), together with the base wine.

The base wine was rated as relatively high in *floral* and *apple*, and relatively low in all other attributes. The juice hydrolysate was significantly more intense in *honey*, *chocolate*, *dried fig* and *tobacco* than the base wine, while the skin hydrolysate was rated as significantly less intense than the base wine in *floral* and *apple*, and

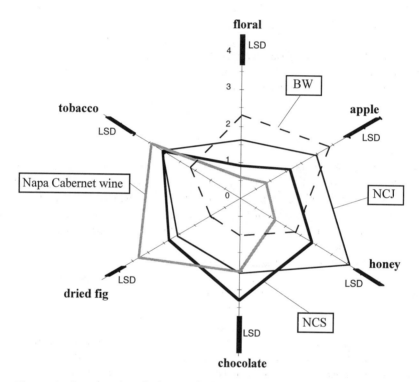

Figure 1. Sensory descriptive analysis data of Napa Cabernet Sauvignon samples and the base wine. Mean ratings of 14 judges x 2 replicates and least significant differences (LSD, p<0.05) are shown. For sample codes, see Table II.

more intense in *chocolate*, *dried fig* and *tobacco*. The three latter attributes were also scored highly for the Cabernet Sauvignon wine sample.

To compare aroma profiles of the 12 glycoside hydrolysate samples, a principal component (PC) analysis of the mean data was performed; the attributes (plotted as vectors) and wine factor scores are plotted for the first two rotated components in Figure 2. The first component contrasted differences in intensity of the samples for the *apple* attribute compared to that of the *tobacco* and *dried fig* attributes. The second component contrasted *chocolate* and *honey* with the *floral* attribute.

The white base wine, to which the glycoside hydrolysates were added, was the most intense in *apple* and *floral*, and was low in all other attributes. Those samples situated furthest from the base wine have the largest difference in aroma produced by the hydrolysates. In general the juice hydrolysates were located closest to the base wine. Thus the skin glycoside hydrolysates were more intense than the juice glycoside hydrolysates in at least one of the other attributes (ie *honey*, *dried fig*, *tobacco* or *chocolate*). This result is of importance because it shows that conventional winemaking practice used for these varieties, ie skin maceration, will be likely to impart flavor to wines due to extraction of glycosides from the skins, followed by hydrolysis upon storage. In noting this effect, it should be recorded that the ratings for the two red wines (see Figure 1 for the Cabernet Sauvignon data, the Merlot wine had the mean scores: *apple* 0.77, *floral* 0.36, *dried fig* 3.3, *chocolate* 2.1, *tobacco* 3.1) were also perceived by the panel to be low in *apple* and *floral* and were high in *dried fig*, *tobacco* and *chocolate*. Thus the skin glycosides gave hydrolysates with aroma properties more similar to that of the wines than the juice glycosides.

In addition to the clear differences that can be seen between the skin and juice glycosides, differences were also apparent due to the other two variables in this experiment, ie grape variety and region of origin. For example, the two Australian skin extract samples were most intense in *tobacco* and *dried fig*, and relatively low in *chocolate*, while the Californian skin extract samples were in general more intense in the *chocolate* attribute. A possible effect of grape variety is illustrated by comparing the Cabernet and Merlot fruit from the same region, with, for example, those from the Davis vineyard exhibiting dissimilar aroma properties. The Cabernet from this vineyard was relatively high in *chocolate* (and *dried fig* and *tobacco* for the skin isolate), while the Merlot was rated as low in all attributes for both skin and juice samples.

The aromas contributed by these black grape glycosides are of interest, as the hydrolysates gave aromas which were unlike that produced from hydrolysis of glycosides isolated from white grapes. In white varieties, attributes such as *lime*, *pineapple* and *toasty* were important to the aroma of the hydrolysates. The attributes *honey* and *tobacco* (as well as the related *tea* attribute) are common in each of the studies carried out *(7-11)*.

In a separate part of this study, the aroma properties of glycoside enzyme hydrolysates added to a white base wine were assessed by duo-trio difference tests with 20 judges. The Australian samples only were evaluated. In tests comparing the

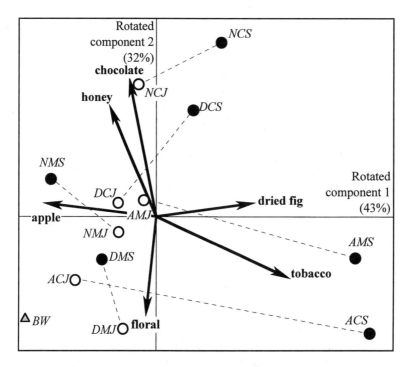

Figure 2. Principal component biplot of rotated components 1 and 2 for mean descriptive analysis ratings (n=14 judges x 2 reps). Vectors for the aroma attributes, and the scores for the fifteen samples are shown. Open symbols indicate juice samples, while closed symbols indicate skin extracts. For sample codes, see Table II.

glycoside concentrate before addition of the enzyme, with the glycoside concentrate after enzyme treatment, it was found that there were no significant differences in aroma for each of the four pairs tested (Australian Cabernet Sauvignon and Merlot, juice and skin extracts). This indicates that glycosidase hydrolysis of the precursors to release intact aglycons is insufficient to produce detectable aroma; acid catalysis is required, to give aroma-active compounds presumably through rearrangement of the aglycons. This was previously suggested by a study on Semillon glycosides *(9)*.

Compositional analyses. The volatile composition of each of the 12 glycoside acid-hydrolysates examined in the sensory descriptive analysis study was analyzed by GC/MS. The enzyme-hydrolysates for the Australian fruit were also subjected to GC/MS analysis. More than 200 compounds were observed and their concentration estimated, and as has previously been reported for Chardonnay, Semillon and Sauvignon Blanc *(14-16)*, almost all could be classed as one of four categories of secondary metabolites: norisoprenoids, benzene derivatives, monoterpenes, and aliphatic compounds. Also, as previously reported, a miscellaneous group of metabolites was found among the glycoside hydrolysates *(16)*. Figure 3 gives the total concentration of the five categories of compounds observed after acid hydrolysis, for the Napa Cabernet Sauvignon and Merlot samples. For both sets of grape samples in Figure 3, and for each of the other grape samples studied (data not shown), the concentration of norisoprenoids and monoterpenes which were investigated in this study was greater in the juices than in the skin extracts. In contrast, the benzenoid class was consistently higher in the skin extracts. The aliphatic class did not show a consistent trend. In comparison to that observed in white varieties and in the variety Shiraz *(12)*, there was a substantially lower concentration of monoterpenes in all of the black grape samples studied here. The dominant class of compounds was the benzene derivatives, and the aliphatic class was also at a relatively high level in these samples compared to that found in earlier studies.

Relationship between the sensory ratings and the volatile composition of the acid hydrolysates. The twelve glycoside acid-hydrolysates differed in aroma and in their volatile composition. To relate the sensory data to the volatile data, the soft modelling technique of Partial Least Squares (PLS) regression analysis was utilized. This procedure [discussed in *(18)*] attempts to account for any common variation between two blocks of data; the compositional data set can be considered in this case as the independent x-block data, with the mean sensory values being the dependent y-block set. The PLS method can be used in situations such as those prevailing in this study, where there are relatively few samples, a large number of x-variables, and where there is substantial noise in the data (error in determination of both descriptive analysis and GC/MS data). From the total number of volatiles quantified, a subset of 53 compounds was included in the PLS analysis. These compounds, together with their codes, and the maximum concentration at which they were observed in any of the hydrolysates, are listed in Table IVa. Also given in Tables IVa and IVb are values for the explained variance from the PLS analysis for each of the attributes and

Table IVa. Compositional variables considered as x-block data in the PLS regression analysis, their codes, maximum value of the 12 samples analyzed, and the percentage explained variance from the first two components extracted from the PLS model.

Compound name (x-data)	Code	max value[a] (µg/L)	Explained variance from PLS model (%)	
			Component 1	Component 2
Aliphatics				
Decanoic acid, ethyl ester	A1	41	0	71
Dodecanoic acid	A2	20	0	0
Dodecanoic acid, ethyl ester	A3	4	10	3
Heptanoic acid	A4	7	78	75
Hexadecanoic acid	A5	64	38	44
Hexadecanoic acid, ethyl ester	A6	41	0	56
Hexanoic acid, ethyl ester	A7	3	0	0
Octadecanoic acid, ethyl ester	A8	6	0	5
Propanedioic acid, diethyl ester	A9	4	0	52
Benzene derivatives				
1,1-Dimethyl ethyl 4-methoxyphenol	B1	31	8	10
2,6-Dimethoxyphenol	B2	24	15	8
4-Hydroxy 3-methoxy benzoic acid, ethyl ester	B3	164	15	7
Acetosyringone	B4	2	8	0
Acetovanillone	B5	6	11	63
Butyrovanillone	B6	27	3	13
cis-cinnamic acid	B7	3	13	75
Ethyl syringate	B8	299	65	69
Ethyl 2-hydroxy phenylpropanoate	B9	32	8	10
unknown benzenoid	B10	4	20	18
p-hydroxy benzoic acid	B11	47	0	0
Propiosyringone	B12	5	0	0
Syringic acid	B13	117	62	62
trans 4-Hydroxy cinnamic acid	B14	9	23	23
Vanillic acid	B15	302	0	2
Vanillin	B16	22	0	92
Ethyl coumarate	B17	5	8	43
unknown methyl ester benzene derivative[b]	B18	10	0	8
Phenol 2,6-methoxy 4-hydroxy	B19	7	50	76
unknown methyl ester methoxy substituted benzene derivative[b]	B20	8	13	36
Monoterpenes				
2,2,6-Trimethyl 6-vinyltetrahydropyran	M1	2	50	44
2,6-dimethyl oct-7-ene, 2,6-diol	M2	11	4	74

Table IVa. *Continued.*

α-Terpineol	M3	14	18	11
cis Chrysanthenol	M4	1	0	0
cis Ocimenol	M5	2	10	12
Furan linalool oxide isomer 1	M6	8	40	73
Geranic acid	M7	2	0	0
trans Chrysanthenol	M8	11	0	0
trans Ocimenol	M9	11	0	74
Benzene methyl (1-methylethenyl)	M10	1	14	38
Norisoprenoids				
2-(3-Hydroxybutenyl)-2,6,6-trimethyl cyclohex-3-ene-1-one	N1	101	0	30
3-Hydroxydamascone	N2	42	0	62
6-Hydroxy 6,7-dihydroedulan	N3	6	3	4
Damascenone	N4	34	0	89
Dehydro-ß-ionone	N5	12	0	82
TDN	N6	13	0	85
Vitispirane	N7	43	0	86
Actinidol 1	N8	32	4	26
Actinidol 2	N9	111	0	67
Others				
2-ethyl 3-methyl maleic anhydride	O1	2	48	65
2-Furan carboxylic acid	O2	18	0	0
Acetyl furan	O3	1	0	60
unknown a	O4	0.6	63	64
unknown b	O5	11	54	68

[a]Mean concentration (n=2). [b]Tentative identification based on interpretation of mass spectral data.

Table IVb. Sensory variables included as y-block data in the PLS regression analysis, maximum rating of the 12 samples analyzed, and the percentage explained variance from the first two components extracted from the PLS model.

Aroma attributes (y-data)	Max rating[a]	Explained variance from PLS model (%)	
		Component 1	Component 2
apple	3.4	71	72
floral	2.0	12	16
honey	3.6	0	66
dried fig	2.3	71	73
chocolate	2.8	28	21
tobacco	3.7	38	31

[a]Mean rating (n=14 judges x 2 replicates).

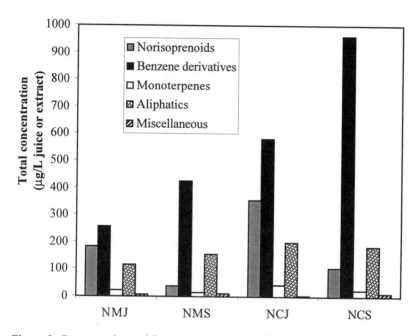

Figure 3. Concentrations of five categories of volatile compounds, observed as a result of acid hydrolysis of the glycoside fractions isolated from juice and skin extracts of Napa Merlot (NMJ, NMS) and Cabernet Sauvignon (NCJ, NCS) fruit.

compounds. If a large proportion of the variance for a variable is explained by one or other of the first two components, then that variable is modelled well by this procedure. Some variables such as provided by the concentration of the compounds A2, A7, A8, B12, B13, B16, B19, M4, M7, M8, N3, and O2 were modelled poorly by the first two components, and are likely to be unrelated to any variation of the sensory attributes of the samples. Additionally the variances of the attributes *floral* and *chocolate* were not explained well by the PLS model for this data set.

PLS extracts components that explain as much of the common variance as possible between the two sets of data. In the present study, the first two components accounted for 36% of the variance of the compositional data, and 46% of the variance in the sensory data. Figure 4a gives the component loadings of both the sensory data and the compositional data, on the first two components. Figure 4b shows the sample scores on these two components. Those compounds, identified by the codes given in Table IV, which are located close to the end of a sensory loading line can be considered to be positively correlated with that sensory attribute. The position of a sample in Figure 4b relative to the position of loadings in Figure 4a indicates the relative importance of the sensory attribute and concentration of volatile compounds to that sample. Thus the Napa Cabernet juice (NCJ) sample is located at the top of Figure 4b, indicating that this sample was scored highest in *honey* (see Figure 4a), and had a relatively high concentration of those compounds closest to the end of the *honey* loading line ie the norisoprenoids N2 (3-hydroxydamascone), N4 (damascenone), N5 (dehydro-ß-ionone), N6 (TDN), N7 (vitispirane), N9 (actinidol isomer), the benzene derivatives B5 (acetovanillone), B8 (cinnamic acid), B17 (vanillin), monoterpenes M2 (an ene diol), M6 (furan linalool oxide isomer), M9 (ocimenol) and the compounds O1, O3, A1 and A9.

The attributes that distinguished several of the skin extract samples, ie *dried fig*, *tobacco* and *chocolate*, appear to be related to numerous compounds, of which B14 (syringic acid), B9 (ethyl syringate), B20 (a methoxy phenol), A4 (heptanoic acid), A5 (hexadecanoic acid), M1, M6, O1, O4 and O5 were most highly correlated. There were few compounds closely linked to the aroma attributes *floral* or *apple*, which is rational as these attributes were important to the base wine used in this study. While it is not possible to determine from this procedure if any of the compounds listed may be actually responsible for particular aroma attributes, the data point to particular compounds which would be worthwhile investigating further by more detailed sensory studies, eg aroma threshold determinations in red wine, and 'GC-sniff' studies.

The glycosyl-glucose (G-G) concentration of the juices and skin extracts. The recent availability of a relatively simple assay to quantify the glycosyl-glucose (G-G) concentration of grapes, juices or wines *(19)* has provided a valuable tool to assess composition in viticulture and winemaking applications. The G-G concentration of a juice or grape sample has been proposed as a possible indicator of wine quality, based on the assumption that it measures the concentration of precursors which can contribute to a wine's sensory properties. A recent study *(9)* showed that for Semillon

26

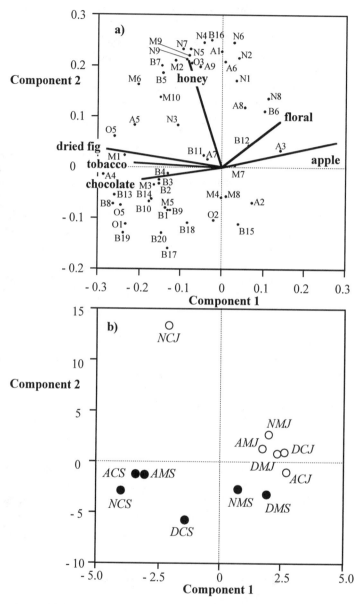

Figure 4. Partial least squares analysis of twelve glycoside hydrolysates, sensory attribute ratings and volatile compound concentration (normalised): a) component loadings, and b) sample scores. For explanation of codes see Tables II and IV.

wines of various ages, there was an inverse linear relationship between the G-G concentration and the score of aroma attributes associated with bottle age. A modification of the assay procedure *(17)* allows the determination of the 'red-free' G-G by subtraction from the measured G-G of the concentration of the glucose moiety of the anthocyanins; the latter is obtained spectrophotometrically. The red-free G-G gives an estimate of the concentration of glycosides other than anthocyanins.

The concentration of total G-G and red-free G-G concentration of the 12 juices and skin extracts was determined, to assess whether there were any relationships among the aroma attributes and these parameters. Linear regressions were performed for each of these variables, against the rotated principal component 1, as discussed above, and represented in Figure 5.

Figure 5 shows a plot of the red-free GG concentration against rotated component 1, and demonstrates that there was a highly significant positive relationship between these two variables. For the rotated component 1 and total G-G there was also a significant relationship (data not shown), but with a substantially lower coefficient of determination (r^2=0.49, p=0.011), while for the color and rotated component 1 there was no statistically significant relationship (r^2=0.11). There was no significant correlation among these variables and any of the other components from the sensory data.

This relationship between G-G and the rotated component 1 indicates that for a juice or skin extract which had a high red-free G-G value, the glycoside fraction isolated from this source was subsequently rated, after acid hydrolysis, as high in those attributes loaded positively on rotated component 1, ie *dried fig* and *tobacco*, and was rated as low in *apple*. It is noteworthy that not all skin extract samples were high in red-free G-G, just as not all juice samples were low. There thus appears to be a good predictive power of the red-free G-G for the intensity of aroma produced from the acid-hydrolysis of glycosides. Correlation between G-G of precursor fractions before acid-hydrolysis and sensory panel score after acid-hydrolysis is consistent with glycoside hydrolysis being responsible for the aroma differences seen.

Conclusion

This investigation has shown that glycosides from Cabernet Sauvignon and Merlot grapes, upon hydrolysis, produced aroma with attributes which were also exhibited by young wines of these varieties. The flavor precursor role of the glycosides of these two varieties is thus indicated. This observation for Cabernet Sauvignon and Merlot adds to comparable findings for Chardonnay, Semillon, Sauvignon Blanc and Shiraz, and may be common for all winemaking varieties.

In red wine production the importance of skin maceration to wine flavor is well known. The effect of skin maceration was assessed in this study by isolating glycosides from either juices or skin extracts. There is also general acceptance by winemakers of the effect of grape origin and grape variety on wine flavor. This investigation has suggested that there was a substantial effect of each of these variables on the aroma released from glycosides. Because these variables were

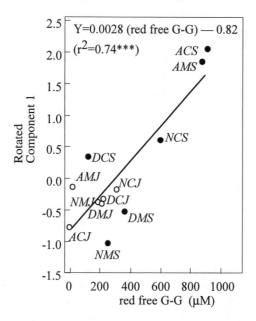

Figure 5. Regression of rotated component 1 of mean aroma attribute scores of glycoside hydrolysates on red free glycosyl-glucose concentration (G-G) of the juices and skin extracts that the glycosides were isolated from. The symbols used are explained in the caption to Figure 2. For sample codes see Table II.

important to the perceived aroma properties of the glycoside samples in this study, the proposition that grape glycosides are of fundamental importance to wine flavor is strengthened. This conclusion is also supported by the relationship found here between the red-free G-G concentration of juice and skin extract samples and their subsequent aroma scores. This finding not only provides further evidence that glycosides contribute aroma for these varieties but supports the proposition that G-G measures on grapes or juices may be a useful and objective means of predicting ultimate strength of wine flavor *(20)*.

Acknowledgments

CA Henschke and Rouge Homme wineries are thanked for providing grape samples. Kevin Scott, David DeSante, Ernie Farinas, and T. Duc Pham are thanked for their assistance. Dr Mark Sefton is thanked for valuable assistance with the interpretation of mass spectra. I.L.F thanks ETS Laboratories for generous support. The Grape and Wine Research and Development Corporation is acknowledged for funding this research.

Literature Cited

1. Williams, P. J.; Allen, M. S. In *Analysis of Fruits and Nuts*; H. F. Linskens and J. F. Jackson, Ed.; Springer-Verlag: Berlin, **1995**; Vol. 18; pp 37-57.
2. Williams, P. J. In *Flavor Science Sensible Principles and Techniques*; T. E. Acree and R. Teranishi, Ed.; American Chemical Society: Washington DC, **1993**; pp 287-303.
3. Stahl-Biskup, E.; Intert, F.; Holthuijzen, J.; Stengele, M.; Schultz, G. *Flavor Fragr. J.* **1993**, *8*, 61-80.
4. Williams, P. J.; Sefton, M. A.; Marinos, V. A. In *Recent Developments in Flavor and Fragrance Chemistry 3rd Haarmann and Reimer International Symposium; 12-15 April, 1992.*; VCH: Kyoto, Japan, 1993; pp 283-290.
5. Winterhalter, P.; Skouroumounis, G. K. *Advances in Biochemical Engineering/Biotechnology* **1997**, *55*, 73-104.
6. Winterhalter, P.; Schreier, P. *Flavor Fragr. J.* **1994**, *9*, 281-287.
7. Francis, I. L.; Sefton, M. A.; Williams, P. J. *J. Sci. Food Agric.* **1992**, *59*, 511-520.
8. Francis, I. L.; Sefton, M. A.; Williams, P. J. *Am. J. Enol. Vitic* **1994**, *45*, 243-251.
9. Francis, I. L.; Tate, M. E.; Williams, P. J. *Aust. J. Grape Wine Res.* **1996**, *2*, 70-76.
10. Williams, P. J.; Francis, I. L. In *Biotechnology for Improved Foods and Flavors ACS Symposium Series 637*; G. R. Takeoka; R. Teranishi; P. J. Williams and A. Kobayashi, Ed.; American Chemical Society: Washington D.C., 1996; pp 124-133.
11. Abbott, N. A.; Coombe, B. G.; Williams, P. J. *Am. J. Enol. Vitic* **1991**, *42*, 167-174.
12. Abbott, N. A. PhD Thesis, University of Adelaide, **1993**.
13. Gomez, E.; Martinez, A.; Laencina, J. *Vitis* **1994**, *33*, 1-4.
14. Sefton, M. A.; Francis, I. L.; Williams, P. J. *Aust. J. Grape Wine Res.* **1996**, *2*, 171-178.
15. Sefton, M. A.; Francis, I. L.; Williams, P. J. *J. Food Sci.* **1994**, *59*, 142-147.

16. Sefton, M. A.; Francis, I. L.; Williams, P. J. *Am. J. Enol. Vitic.* **1993**, *44*, 359-370.
17. Iland, P. G.; Cynkar, W.; Francis, I. L.; Williams, P. J.; Coombe, B. G. *Aust. J. Grape Wine Res.* **1996**, *2*, 171-178.
18. Chien, M.; Peppard, T. In *Flavor Measurement*; C.-T. Ho and C. H. Manley, Ed.; Marcel Dekker, Inc.: New York, 1992; pp 1-35.
19. Williams, P. J.; Cynkar, W.; Francis, I. L.; Gray, J. D.; Iland, P.; Coombe, B. G. *J. Agric. Food Chem.* **1995**, *43*, 121-128.
20. Abbott, N. A.; Williams, P. J.; Coombe, B. C. In *Proceedings of the Eighth Australian Wine Industry Technical Conference*; Winetitles: Melbourne, Victoria, 1992; pp 72-75.

Chapter 3

Methoxypyrazines of Grapes and Wines

M. S. Allen[1] and M. J. Lacey[2]

[1]National Wine and Grape Industry Centre, Charles Sturt University, P.O. Box 588, Wagga Wagga, NSW 2678, Australia
[2]Division of Entomology, CSIRO, G.P.O. Box 1700, Canberra, ACT 2601, Australia

Methoxypyrazines are grape-derived compounds that contribute vegetative/herbaceous aroma to Sauvignon blanc, Semillon and Cabernet Sauvignon wines. Three methoxypyrazines have been identified, isobutylmethoxypyrazine is predominant, and they have structures consistent with a related biosynthetic origin. They occur at trace levels, with a combined concentration of, typically, 1-40 ng/L. There is a narrow concentration window that allows their flavor contribution to be evident yet not excessive. The concentration of these methoxypyrazines in grapes, and their impact in the resulting wines, is strongly and systematically influenced by viticultural conditions, such as the temperature during ripening, the berry maturity, and the fruit exposure to sunlight. Ethylmethoxypyrazine has also been identified in grapes and wines. A different biosynthetic origin is suggested by its structure and by a lack of dependence of its occurrence on viticultural conditions.

Methoxypyrazines are grape-derived flavor compounds that contribute a very characteristic vegetative, herbaceous, bell pepper or earthy aroma to wines of some grape varieties. Three methoxypyrazines, 2-methoxy-3-(2-methylpropyl)pyrazine (**1**) (isobutylmethoxypyrazine), 2-methoxy-3-(1-methylpropyl)pyrazine (**2**) (*sec*-butylmethoxypyrazine) and 2-methoxy-3-(1-methylethyl)pyrazine (**3**) (isopropylmethoxypyrazine) have been found to contribute such aroma. All three have extremely low sensory detection thresholds of 1-2 ng/L in water (*1, 2, 3*), so even ultra-trace concentrations of these compounds in grapes can have a marked impact on the resulting wine flavor. In some winemaking regions, their distinctive aroma is considered important to the regional style of Sauvignon blanc wines; in other regions, their aroma is disliked. At low concentrations within the range of their occurrence, methoxypyrazines provide aroma that appears to be important in distinguishing wines of the grape varieties Sauvignon blanc and Cabernet Sauvignon from wines of other grape varieties. However, at high concentration within the range of their occurrence, their aroma can be overpowering and unpleasant. Fortunately, their occurrence has always shown a clear and consistent relationship to the grape variety and to the conditions under which the vine is grown. Study of these influences is helping our efforts to consistently produce grapes of the quality desired for high quality wine production.

$$\text{N} \underset{\text{N}}{\overset{\text{R}}{\bigcirc}} \text{OCH}_3 \qquad\qquad \text{N} \underset{\text{N}}{\overset{\text{R}}{\bigcirc}} \text{OCD}_3$$

1 $R = CH_2CH(CH_3)_2$ **5** $R = CH_2CH(CH_3)_2$
2 $R = CH(CH_3)CH_2CH_3$
3 $R = CH(CH_3)_2$ **6** $R = CH(CH_3)_2$
4 $R = CH_2CH_3$

Methoxypyrazines of Flavor Importance in Wine

Quantitative Analysis. To study those factors that influence the occurrence of methoxypyrazines, it was necessary to be able to quantify them. Furthermore, it was desirable that this be possible even at the lowest levels that might be relevant. This indicated a need for their quantitative analysis to extend to concentrations below their sensory detection threshold of 1-2 ng/L, preferably by an order of magnitude. Accurate quantitative analysis at such extremely low analyte concentrations is difficult to achieve. If a moderate sample size is to be used, and if the analysis is to cope with some losses in isolation, then the technique needs to have a detection limit of a few picograms. Furthermore, it must provide quantitative data with adequate accuracy and precision for meaningful interpretation.

For rigorous quantitative analysis at such levels, stable isotope dilution mass spectrometry is clearly the method of choice. In this technique, the internal standard is exactly the same component as the analyte except that the internal standard contains an isotopic label that distinguishes it from the natural material. This ensures that the analyte and internal standard behave identically, and that both are isolated and measured with identical efficiency. At trace levels this is important, as slight differences of chemical behavior, volatility or detection efficiency between the internal standard and the analyte can lead to very significant quantitation errors. This stable isotope dilution method was developed for methoxypyrazines (**1**) and (**3**) by synthesizing the trideuterated methoxypyrazines (**5**) and (**6**) as internal standards. Analysis uses gas chromatography-mass spectrometry (GC-MS). Selected ion monitoring improves sensitivity and selectivity, and positive ion ammonia chemical ionization provides ionization selectivity. A detection limit of about 200 femtograms can be achieved (*4, 5*), allowing quantitative analysis of methoxypyrazines to well below the sensory detection threshold, typically with a detection limit as low as 0.1 ng/L and a limit of quantitation of 0.3 ng/L.

In studies of methoxypyrazines, a stable isotope labeled internal standard has not always been used. Calo *et al.* (*6*) used *sec*-butylmethoxypyrazine as an internal standard to quantify isobutylmethoxypyrazine in a comparison of grape varieties. Recently, Hashizume and Umeda have used 2-methyl-3-*n*-propylpyrazine as an internal standard to quantify methoxypyrazines in Japanese red wine and grape samples (*7*). However, the increased potential for lack of precision and accuracy needs to be recognized, and the natural occurrence of *sec*-butylmethoxypyrazine is a drawback to its use as an internal standard.

The difficulty of achieving precision in quantitative analysis at trace levels is highlighted by a long-term study that we have made of the mass spectrometric determination of methoxypyrazines (**1**), (**2**) and (**3**) in a prepared standard mixture, using the trideuterated isobutylmethoxypyrazine (**5**) as internal standard (Table I). With 200 pg injections, a level that corresponds to the analysis of a wine with a relatively high methoxypyrazine concentration, there was a significantly increased coefficient of variation for the determination of the concentration of (**2**) by comparison with determination of (**1**). This shows the extent to which the precision of determination can be degraded by even the minor structural difference introduced by positional isomerism in the C_4 alkyl side chain. The effect is even more marked when the analyte, as in (**3**), possesses one carbon atom less than the internal standard. For this reason, virtually all our work has been performed with two stable isotope labeled internal standards (**5**) and (**6**), to ensure accuracy of determination of the naturally occurring methoxypyrazines (**1**) and (**3**). For determination of (**2**), the closely related internal standard (**5**) is used, and the expectation of a *ca.* 10% error in the determination of (**2**) is accepted.

Table I.   Coefficient of variation of replicate mass spectrometry determinations of methoxypyrazines (1), (2) and (3) using trideuterated isobutylmethoxypyrazine (5) as internal standard

Methoxypyrazine analyte	%CV[1]
Isobutylmethoxypyrazine (**1**)	3.7
sec-Butylmethoxypyrazine (**2**)	10.5
Isopropylmethoxypyrazine (**3**)	21.2

[1]Analysed over an 18 month time period; n = 36.

Compounds of Importance. The three identified methoxypyrazines (**1**), (**2**) and (**3**) are very similar, both in structure and in the factors that influence their occurrence. One of them, isobutylmethoxypyrazine (**1**), is almost invariably dominant; typically, it has an eight-fold or higher concentration than the other methoxypyrazines (Figure 1). Of the other two methoxypyrazines, the most abundant is isopropylmethoxypyrazine (**3**). Its concentration sometimes exceeds its sensory detection threshold. As they all have a similar sensory detection threshold, of 1-2 ng/L in water, isobutylmethoxypyrazine (**1**) will be the principle contributor to aroma. However, some evidence suggests that isopropylmethoxypyrazine in red wine may be more important than its sensory detection threshold suggests (*8*); it may act synergistically with isobutylmethoxypyrazine, with its more earthy aroma (*8, 9, 10, 11*) slightly modifying the overall perceived flavor. In one wine, isopropylmethoxypyrazine was as abundant as isobutylmethoxypyrazine, allowing confirmation of the occurrence of isopropylmethoxypyrazine by recording its full-scan mass spectrum (*12*).

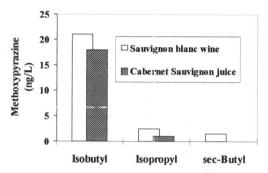

Figure 1. Typical concentrations of methoxypyrazines (**1**), (**2**) and (**3**) (Adapted from ref. 13. Copyright 1996 Winetitles).

The co-occurrence of these three methoxypyrazines is consistent with a bio-synthetic pathway (Figure 2) proposed over 20 years ago (*14*). The amino acid leucine is envisaged as the source of the C_4 side chain of the methoxypyrazine, through condensation of its amino amide with an unspecified C_2 component, and methylation of the initial pyrazinone condensation product. This proposed biosynthetic pathway readily accommodates all three methoxypyrazines through incorporation of either leucine, isoleucine or valine, all of which are commonly available amino acids in plants. Although the validity of this pathway in vines or other plant material is unknown, the major features of this proposed pathway have been shown to apply to the biosynthesis of isopropylmethoxypyrazine by certain bacteria (*15, 16*).

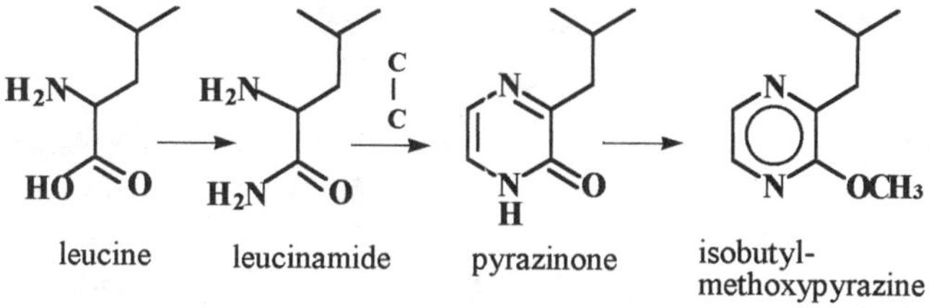

Figure 2. Biosynthetic pathway to isobutylmethoxypyrazine proposed by Murray and Whitfield (*14*).

Factors Affecting their Concentration in Wine. Studies of the occurrence of isobutylmethoxypyrazine (**1**) have shown that it has a consistent and systematic relationship to the grape variety and the vine growing conditions.

In winemaking, the occurrence of methoxypyrazine-like aroma is consistently related to the grape variety. So there can be little doubt that these methoxypyrazines are produced under genetic control in the grape berry. Analysis confirms this, for while Cabernet Sauvignon, Sauvignon blanc and Semillon produce significant isobutyl-methoxypyrazine levels, some other cultivars do not seem to produce this compound at all. In a comparative study of different grape varieties, using vines within the same vineyard, high levels of isobutylmethoxypyrazine were evident in Cabernet Sauvignon and Sauvignon blanc grapes, but Pinot noir showed no detectable methoxypyrazine at ether véraison or normal harvesting maturity (Allen, M.S., Charles Sturt University, unpublished data).

With increasing grape berry maturity there is a profound decrease in the concentration of isobutylmethoxypyrazine (**1**) (Table II). Comparatively high levels, often over 100 ng/L, are present at véraison in the fruit of Cabernet Sauvignon (*17*) and Sauvignon blanc (*5*) grape varieties. However, these levels fall very rapidly in the early stages of ripening, and they can be less than 1% of the véraison concentration by the time of harvesting. Ripening temperature also has an impact. At comparable stages of fruit ripeness, substantially higher methoxypyrazine levels occur in cool regions by comparison with warm regions (*5*, *18*). In warm areas, the level of isobutylmethoxypyrazine can fall well below its sensory detection threshold by the time of fruit harvesting, whereas in cool areas it can be 20-30 ng/L.

During fermentation with grape skin contact, we have consistently found an increase of the concentration of isobutylmethoxypyrazine, a situation that suggests that methoxypyrazines may either be extracted from the solid parts of the grape or be produced by yeast-mediated effects. Contact of the juice with the grape skins is required for this increase to occur, and the increase is slow, following the progress of fermentation (*19*).

The influence of the vine canopy and the pruning and training system can also be very important. Particularly as fruit exposure to light influences the methoxypyrazine level significantly. Within the vine canopy, the more exposed fruit provides a consistently lower level of methoxypyrazines than the more shaded fruit, typically half or less of the level of that in the most shaded fruit within the canopy (*20*).

Desired Concentration in Wine. Recognition of the character of methoxypyrazine aroma as 'herbaceous'or 'vegetative' occurred at 4-8 ng/L in white wine, but it is clear from work with Sauvignon blanc wines that 30 ng/L is often considered to be overpowering and out of balance (*3*). This indicates a surprisingly narrow concentration window for methoxypyrazines if they are to contribute usefully to wine flavor. The desirable range in Sauvignon blanc wines appears to be 8-15 ng/L. A similar situation is indicated for Cabernet Sauvignon wines. Analysis of a range of Cabernet Sauvignon-based red wines has indicated a concentration range of isobutylmethoxypyrazine of 7-15 ng/L in Bordeaux wines with good flavor balance and 27-29 ng/L in some wines that were showing distinct methoxypyrazine aroma (*18*). It is possible that the greater flavor complexity and intensity of some red wines may mask methoxypyrazine aroma to some degree, permitting the presence of higher levels of methoxypyrazines in those wines than in wines with less intense and less complex flavor. Curiously, a study of the perception of added methoxypyrazines to a red wine (*8*) found a higher aroma threshold for isobutylmethoxypyrazine (**1**) than for iso-propylmethoxypyrazine (**3**). This suggests that low levels of isopropylmethoxypyrazine may be more important to the perception of red wines than is indicated by the sensory detection threshold in water.

Ethylmethoxypyrazine

In 1982, tentative evidence for the occurrence of ethylmethoxypyrazine (**4**) in grape juice was reported (*21*). The methoxypyrazine had, at that time, been identified in potato also (*22, 23*) and studies with synthetic ethylmethoxypyrazine had shown that it has a more earthy, potato-like aroma than isobutylmethoxypyrazine. Its sensory detection threshold of 425 ng/L (*2*) is much higher than that of isobutylmethoxy-pyrazine. For this reason, interest in its possible occurrence was initially limited by the expectation that, if it occurred at concentrations typical of isobutylmethoxypyrazine, it would not have an effect on flavor. The natural occurrence of ethylmethoxypyrazine (**4**) had not always been verified (*10*). Evidence for its occurrence was weak, and a consistent difficulty in those studies had been a lack of enough material to record a full-scan mass spectrum. A further concern was that the biosynthetic pathway that had been proposed for methoxypyrazines (*14*), in which an amino acid is the source of the alkyl side chain (Figure 2), would require an unusual amino acid, 2-aminobutyric acid, for ethylmethoxypyrazine formation.

Occurrence in Wine. The occurrence of ethylmethoxypyrazine (**4**) in wine has now been definitively confirmed, and we have identified this methoxypyrazine in grape juice and have studied factors that might influence its occurrence (Allen, M. S., Boyd, S. J., in preparation). Unlike isobutylmethoxypyrazine, the concentration of ethylmethoxy-pyrazine in wine was found to be highly variable. Indeed, the concentration of ethyl-methoxypyrazine in two wines was so high that it allowed verification of the structure of this methoxypyrazine by comparison of its full-scan mass spectrum and its gas chro-matography retention time, on several stationary phases of widely differing polarity, with that of synthetic material. Over 100 ng/L was found in a Pinot noir wine that contained no detectable isobutylmethoxypyrazine. Furthermore, a concentration of 1000 ng/L, a level well above its sensory detection threshold of 425 ng/L, was found in a Cabernet Sauvignon wine that contained about 10 ng/L of isobutyl-methoxypyrazine. In the latter wine, ethylmethoxypyrazine can clearly be expected to contribute to the wine's sensory character.

Response to Viticultural Conditions. In grapes, the behavior of ethylmethoxy-pyrazine (**4**) is quite different to that of isobutylmethoxypyrazine (**1**). Comparison of the concentration of these two methoxypyrazines during ripening provides an example.

Isobutylmethoxypyrazine decreased consistently, but ethylmethoxypyrazine displayed a concentration that was somewhat erratic (Table II). A similar situation was found when the effect of light interception by the fruit was examined. Higher light levels decreased the concentration of isobutylmethoxypyrazine but did not consistently influence ethylmethoxypyrazine.

Table II. Effect of Ripening on Grape Methoxypyrazine Concentration for two Vine Pruning Systems[a]

Date[b]	Isobutylmethoxypyrazine (ng/L)		Ethylmethoxypyrazine (ng/L)	
	Minimal pruning	Spur pruning	Minimal pruning	Spur pruning
Jan.25	111.0	188.5	21.2	13.9
Feb.1	63.1	122.6	2.3	38.0
Feb.8	45.6	90.7	1.4	4.9
Feb.24	11.1	18.3	4.1	2.9
Mar.3	10.5	16.3	3.2	3.0
Mar.9	6.5	9.9	2.4	47.1

[a]Concentration in freshly extracted juice of Cabernet Sauvignon grapes.
[b]Dates are southern hemisphere growing season.
Reprinted with permission from ref. 24

Origin. Although ethylmethoxypyrazine is clearly identified as a component of grapes and wines, its origin is uncertain. It does not show the clear relationship to vine variety, berry development and canopy light penetration that is found with isobutylmethoxypyrazine, and it does not comfortably fit the biosynthetic pathway that appears likely for the other methoxypyrazines. The evidence strongly implicates that ethylmethoxypyrazine has a different origin to isobutylmethoxypyrazine. It may have a plant origin, but equally a microbial origin cannot be excluded. It may also arise from a precursor or as an artifact of the isolation conditions. Furthermore, there may even be a difference in origin between the low levels usually found in grapes and wines and the comparatively high level found occasionally in wine. Although it is usually well below its sensory detection threshold, it can occasionally occur in wine at much higher concentration and has, in one case, well exceed its sensory detection threshold.

Acknowledgments

We thank Stephen Boyd for mass spectrometry analysis. Financial support was provided by the Grape and Wine Research Council and by the Grape and Wine Research and Development Corporation.

38

Literature Cited

1. Buttery, R. G.; Seifert, R. M.; Lundin, R. E.; Guadagni, D. G.; Ling, L. C. *Chem. Ind.* (London) **1969**, 490-491.
2. Seifert, R. M.; Buttery, R. G.; Guadagni, D. G.; Black, D. R.; Harris, J. G. *J. Agric. Food Chem.* **1970**, *18*, 246-249.
3. Allen, M. S.; Lacey, M. J.; Harris, R. L. N.; Brown, W. V. *Am. J. Enol. Vitic.* **1991**, *42*,109-112.
4. Harris, R. L. N.; Lacey, M. J.; Brown, W. V.; Allen, M. S. *Vitis* **1987**, *26*, 201-207.
5. Lacey, M. J.; Allen, M. S.; Harris, R. L. N.; Brown, W. V. *Am. J. Enol. Vitic.* **1991**, *42*, 103-108.
6. Calo, A.; Di Stefano, R.; Costacurta, A; Calo, G. *Riv. Viticult. Enol.* **1991**, *44*, 3-25.
7. Hashizume, K.; Umeda, N. *Biosci. Biotech. Biochem.* **1996**, *60*, 802-805.
8. Maga, J. A. In *Flavors and Off-Flavors, Proceedings of the 6th International Flavor Conference;* Charalambous, G., Ed.; Elsevier: Amsterdam, 1990; pp 61-70.
9. Murray, K. E.; Shipton, J.; Whitfield, F. B. *Chem. Ind.* (London) **1970**, 897-898.
10. Buttery, R. G.; Ling, L. C. *J. Agric. Food Chem.* **1973**, *21*, 745-746.
11. Parliment, T. H.; Epstein, M. F. *J. Agric. Food Chem.* **1973**, *21*, 714-716.
12. Allen, M. S.; Lacey, M. J.; Boyd, S. *J. Agric. Food Chem.* **1995**, *43*, 764-772.
13. Allen, M. S.; Lacey, M. J.; Boyd, S. J. In *Proceedings of the Ninth Australian Wine Industry Technical Conference;* Stockley, C. S., Sas, A. N., Johnstone, R. S., Lee, T.H., Eds.; Winetitles: Adelaide, 1996; pp 83-86.
14. Murray, K. E.; Whitfield, F. B. *J. Sci. Food Agric.* **1975**, *26*, 973-986.
15. Gallois, A.; Kergomard, A.; Adda, J. *Food Chem.* **1988**, *28*, 299-309.
16. Leete, E.; Bjorklund, J. A.; Reineccius, G. A.; Cheng, T.-B. In *Bioformation of Flavours;* Patterson, R. L. S.; Charlwood, B. V.; MacLeod, G.; Williams, A. A., Eds.; Royal Society of Chemistry: Cambridge, UK, 1992; pp 75-95.
17. Allen, M. S.; Lacey, M. J.; Harris, R. L. N.; Brown, W. V. *Aust. N. Z. Wine Ind. J.* **1990**, *5*, 44-46.
18. Allen, M. S.; Lacey, M. J.; Boyd, S. *J. Agric. Food Chem.* **1994**, *42*, 1734-1738.
19. Allen, M. S. *Aust. Grapegrower and Winemaker* **1994** (366), 22-23.
20. Allen, M. S.; Lacey, M. J. *Vitic. Oenol. Sci.* **1993**, *48*, 211-213.
21. Augustyn, O. P. H.; Rapp, A.; van Wyk, C. J. *S. Afr. J. Enol. Vitic.* **1982**, *3*, 53-60.
22. Meigh, D. F.; Filmer, A. A. E.; Self, R. *Phytochemistry* **1973**, *12*, 987-993.
23. Nursten, H. E.; Sheen, M. R. *J. Sci. Food Agric.* **1974**, *25*, 643-663.
24. Allen, M. S.; Lacey, M. J.; Boyd, S. J. In *Biotechnology for Improved Foods and Flavors;* Takeoka, G. R., Teranishi, R., Williams, P. J.; Kobayashi, A. Eds; ACS Symposium Series 637; American Chemical Society: Washington, DC., 1996; pp 220-227.

Chapter 4

Comparison of Different White Wine Varieties in Odor Profiles by Instrumental Analysis and Sensory Studies

H. Guth

Deutsche Forschungsanstalt für Lebensmittelchemie, Lichtenbergstrasse 4, 85748 Garching, Germany

Two different white wine varieties (Gewürztraminer and Scheurebe), which differ in their odor profiles, were investigated by gas chromatography-olfactometry (GC-O). Aroma extract dilution analysis (AEDA) and static headspace analysis-olfactometry (SHA-O) yielded 41 and 45 odor-active compounds for Scheurebe and Gewürztraminer wines, respectively. An unknown compound with coconut-like and woody odor qualities, which has not yet been detected in wine or a food, was identified as (3S,3aS,7aR)-3a,4,5,7a-tetrahydro-3,6-dimethylbenzofuran-2(3H)-one (wine lactone). Quantitation and calculation of odor activity values of potent odorants showed, that differences in odor profiles of both varieties were mainly caused by cis-rose oxide in Gewürztraminer and 4-mercapto-4-methyl-pentan-2-one in Scheurebe. Reconstruction of the flavor and quantitation of potent odorants in the different stages of wine making of Gewürztraminer will be discussed.

Up to now more than 680 volatile compounds have been identified in different white wine varieties (*1*) but little is known about the actual contribution to the overall flavor. This paper summarizes the screening experiments of the most odor active compounds in Gewürztraminer and Scheurebe wines by aroma extract dilution analysis (AEDA) and static headspace analysis-olfactometry (SHA-O), followed by quantitation and calculation of odor activity values (OAV's). Reconstruction of the flavor of both varieties and sensory studies will be discussed. Furthermore the influence of various ethanol concentrations on the overall flavor profile of Gewürztraminer wine was examined. Investigations about changes during the different stages of wine making of Gewürztraminer (after pressing of grapes, after yeast fermentation, after malolactic fermentation and after maturing in high-grade steel tank) will be reported. The influence of barrel aging on the overall flavor of Gewürztraminer wine will be the subject of later sections of the present paper.

Gas chromatography-olfactometry (GC-O)

The odorants, which contribute significantly to the flavor of a food, can be localized in the capillary gas chromatogram of the volatile fraction by gas chromatography-olfactometry (GC-O) (*2, 3*). Various methods were developed to determine the odor-activity of the eluting compounds. Using Charm-analysis Chrisholm et al. (*4*), Schlich and Moio (*5*) and Moio et al. (*6*) evaluated ß-damascenone, 3-methylbutyl acetate, 2-phenylethanol, vanillin, butan-2,3-dione, guaiacol, 4-vinylguaiacol, ethyl cinnamate, linalool and various ethylesters, as the most potent odorants of Chardonnay and White Riesling wines. By application of the Osme technique, Miranda-Lopez et al. (*7*) investigated the volatile fractions of different vintages of the variety Pinot noir. High Osme values were found for 3-methylbutanol, 2-phenylethanol, 2-phenylethyl acetate, hexanoic acid, γ-nonalactone and 3-(methylthio)-1-propanol. Berger (*8*) identified (E)-ß-damascenone and phenylethanol as key odorants of Chardonnay-Semillon wines, as these compounds showed the highest flavor dilution (FD)- factors in aroma extract dilution analysis (AEDA).

Recently, AEDA and SHA-O yielded 41 and 45 odor active compounds for Scheurebe and Gewürztraminer wines, respectively (*9*). Ethyl 2-methylbutyrate, ethyl isobutyrate, 2-phenylethanol, 3-methylbutanol, 3-hydroxy-4,5-dimethyl-2(5H)-furanone, 3-ethylphenol and one unknown compound, named wine lactone, showed high flavor dilution (FD)- factors (Table I) in Gewürztraminer and Scheurebe wines. 4-Mercapto-4-methylpentan-2-one belongs to the most potent odorants only in the variety Scheurebe whereas cis-rose oxide was perceived only in Gewürztraminer (Table I). 4-Mercapto-4-methylpentan-2-one was identified for the first time in Sauvignon blanc wines (*10*). The unknown compound with coconut, woody and sweet odor quality, which has not yet been detected in wine or a food, was identified as 3a,4,5,7a-tetrahydro-3,6-dimethylbenzofuran-2(3H)-one (wine lactone) (*11*).

Because of the three asymmetric centers in the molecule there exist eight different stereoisomers. To identify the stereochemistry of wine lactone syntheses for the enantiomers were developed. On the basis of enantioselective gas chromatography the stereochemistry of wine lactone was in agreement with the 3S,3aS,7aR-enantiomer (*12*); for this stereoisomer a low odor threshold was determined (0.00002 ng/L air):

(3S,3aS,7aR)- 3a,4,5,7a-Tetrahydro-3,6-dimethylbenzofuran-2(3H)-one

Table I. Results of Aroma Extract Dilution Analysis of Gewürztraminer and Scheurebe

Compound	FD-Factor	
	Scheurebe	Gewürztraminer
Wine lactone	1000	1000
Ethyl isobutyrate	100	10
Ethyl 2-methylbutyrate	100	100
3-Methylbutanol	100	100
2-Phenylethanol	100	100
3-Ethylphenol	100	100
3-Hydroxy-4,5-dimethyl-2(5H)-furanone	100	100
Ethyl 3-methylbutyrate	10	10
Ethyl butyrate	10	10
2-Methylpropanol	10	10
Ethyl hexanoate	10	10
cis-Rose oxide	<1	10
4-Mercapto-4-methylpentan-2-one	10	<1
Ethyl octanoate	10	10
Acetic acid	10	10
Linalool	10	10
Butyric acid	10	10
2-/3-Methylbutyric acid	10	10
5-Ethyl-4-hydroxy-2-methyl-3(2H)-furanone	10	10
Ethyl trans-cinnamate	10	10

Source: Data are from ref. *11*.

A dilution experiment by SHA-O indicated acetaldehyde, ethyl acetate, dimethylsulfide and dimethyltrisulfide as further potent odorants in Scheurebe and Gewürztraminer wines. AEDA and SHA-O yielded the same assessment of 4-mercapto-4-methylpentan-2-one and cis-rose oxide, which are responsible for the odor difference of the two varieties, investigated in this study. It should be mentioned that only one sample of each variety was analyzed and for more generality further investigations are necessary.

Quantitation and Calculation of Odor Activity Values

AEDA and SHA-O are suitable tools for recognition of odor active compounds (*13, 14*), but the methods are afflicted with some simplifications: no corrections were made for the losses of odorants during isolation procedure. By AEDA the complete amounts of the odorants present in the solvent extracts are volatilized during GC-O and therefore ranked according to their odor thresholds in air, but the contribution of an odorant to the overall flavor in a food is strongly affected by its odor threshold in the food matrix. Odor thresholds in air are generally much lower than those in water/ethanol mixtures, e.g. for wine lactone an odor threshold of 0.00002 ng/L was found in air, whereas in water/ethanol (9 + 1, w/w) a value of 0.01 µg/L was obtained.

To establish exactly the flavor differences between Scheurebe and Gewürztraminer wines, it is therefore necessary to quantify the levels of recognized odorants and to calculate the odor activity values (OAV's). According to Rothe and Thomas (15) the OAV is defined as ratio of concentration to odor threshold value of the compound.

A suitable tool for the quantitation of trace compounds in foods is a stable isotope dilution assay (IVA) (16, 17). Allen et al. (18) used the IVA for the quantification of two methoxypyrazines in red wines, Guth (11) quantified wine lactone in various red and white wines and Aubry et al. (19) used the technique for the determination of four esters (ethyl dihydrocinnamate, ethyl cinnamate, methyl anthranilate and ethyl anthranilate) in Pinot Noir wines.

42 wine odorants, identified by AEDA and SHA-O (9) in Gewürztraminer and Scheurebe, were quantified by IVA or by using similar internal standards (20). The amounts of potent odorants found in the varieties Scheurebe and Gewürztraminer are listed in Table II. Differences between the two varieties were found for ethyl isobutyrate, which was higher in Scheurebe wine (480 µg/L) than in Gewürztraminer (150 µg/L), whereas cis-rose oxide predominated in the latter, with 21 µg/L compared to 3.0 µg/L in the former wine. Another significant difference was found for 4-mercapto-4-methylpentan-2-one, that was present only in the variety Scheurebe (0.4 µg/L) but not in Gewürztraminer (<0.01 µg/L). Our results are in agreement with investigations of Schreier et al. (21), who found also higher amounts of rose oxide in the variety Gewürztraminer than in Scheurebe.

To estimate the sensory contribution of the 42 odorants to the overall flavor of the wine samples, their OAV's were calculated (Table II). To take into account the influence of ethanol, the odor threshold values of wine odorants were determined in a mixture of water/ethanol (9 + 1, w/w) and were used to calculate the OAV's for each compound. According to the results in Table II, 4-mercapto-4-methylpentan-2-one, ethyl octanoate, ethyl hexanoate, 3-methylbutyl acetate, ethyl isobutyrate, (E)-ß-damascenone, linalool, cis rose oxide and wine lactone showed the highest OAV's in the Scheurebe wine. With exception of 4-mercapto-4-methylpentan-2-one the above mentioned odorants also showed the highest OAV's in Gewürztraminer wine. Differences in the OAV's of ethyl octanoate, ethyl hexanoate, 3-methylbutyl acetate and ethyl isobutyrate between the two varieties are probably caused by differences in the maturity of the fruit at harvest and/or by the fermentation process.

Calculation of OAV's indicated that significant differences in odor profiles of both varieties, investigated in this study, were mainly caused by cis-rose oxide in Gewürztraminer and 4-mercapto-4-methylpentan-2-one in Scheurebe. Investigations about the formation of 4-mercapto-4-methylpentan-2-one in wine were performed by Tominaga et al. (22), who found that the compound was released from an odorless must extract by a cysteine-ß-lyase. The authors suggested that the compound was bound in form of S-(4-methylpentan-2-one)-l-cysteine in grape must.

**Table II.   Concentrations and Odor Activity Values (OAV´s ≥ 10) of Potent
Odorants of Scheurebe and Gewürztraminer Wines**

	Concentration (µg/L)			
Odorant	Scheurebe		Gewürztraminer	
4-Mercapto-4-methylpentan-2-one	0.40	(667)[a]	<0.01	(<1)[a]
Ethyl octanoate	270	(135)	630	(315)
Ethyl hexanoate	280	(56)	490	(98)
3-Methylbutyl acetate	1450	(48)	2900	(97)
Ethyl isobutyrate	480	(32)	150	(10)
(E)-ß-Damascenone	0.98	(20)	0.84	(17)
Linalool	307	(20)	175	(12)
cis-Rose oxide	3.0	(15)	21	(105)
Wine lactone	0.10	(10)	0.10	(10)
Ethyl butyrate	184	(9)	210	(11)

[a]The odor activity values (OAV´s) were calculated as the ratio of concentration to odor threshold value of the compound in water/ethanol (9 + 1, w/w).
Source: Data are from ref. *20*.

Sensory Experiments

Reconstruction of the flavor of both varieties and sensory studies should show, whether the odorants which were detected by AEDA and SHA-O and then quantified, represent the characteristic flavor of Scheurebe and Gewürztraminer wines. Therefore the 42 odorants quantified in Gewürztraminer and Scheurebe wines, respectively, were dissolved in water/ethanol (9 + 1, w/w) and the resulting model mixtures were compared nasally with the original wines (*20*). The model mixtures showed good agreement with the original wines of Gewürztraminer and Scheurebe, respectively.

To clarify, whether the odorants showing high OAV´s are actually the key aroma compounds of Gewürztraminer and Scheurebe, 42 model mixtures were prepared, in which one odorant at a time was omitted (*20*). The absence of cis-rose oxide diminished strongly the similarity of the model mixture with that of the original Gewürztraminer wine. Also the respective omission of wine lactone, ethyl octanoate, acetic acid, 3-methylbutyl acetate, ethyl hexanoate, acetaldehyde, 3-hydroxy-4,5-dimethyl-2(5H)-furanone, geraniol and (E)-ß-damascenone led to a decreasing similarity with the original Gewürztraminer. The respective absence of the remaining 32 compounds was not noticed by the assessors. For Scheurebe model the lack of 4-mercapto-4-methylpentan-2-one had a drastic effect, because the mixture of the remaining 41 odorants showed an odor profile strongly different to the original wine. The ten most potent odorants from the above mentioned sensory experiment were combined to a model mixture (model A, Table III). The aroma of model A was different (similarity 1.5) from that of the original wine. An improvement was achieved in model B

Table III. **Similarity of Various Gewürztraminer Models (A-C)[a] with the Original Wine**

Compound	A	B	C
Acetaldehyde	+	-	+
Ethyl hexanoate	+	+	+
cis-Rose oxide	+	+	+
Ethyl octanoate	+	+	+
Acetic acid	+	-	+
(E)-ß-Damascenone	+	+	+
Geraniol	+	-	+
4,5-Dimethyl-3-hydroxy-2(5H)-furanone	+	-	+
Wine lactone	+	+	+
Ethyl isobutyrate	-	+	+
Ethyl butyrate	-	+	+
Linalool	-	+	+
Ethyl acetate	-	-	+
1,1-Diethoxy ethane	-	-	+
Butan-2,3-dione	-	-	+
Ethyl 2-methylbutyrate	-	-	+
Ethyl 3-methylbutyrate	-	-	+
2-Methylpropanol	-	-	+
3-Methylbutanol	-	-	+
Dimethyltrisulfide	-	-	+
(3-Methylthio)-1-propanol (methionol)	-	-	+
Hexanoic acid	-	-	+
2-Phenylethanol	-	-	+
trans-Ethyl cinnamate	-	-	+
4-Allyl-2-methoxyphenol (eugenol)	-	-	+
(Z)-6-Dodecenoic acid-γ-lactone	-	-	+
4-Hydroxy-3-methoxybenzaldehyde	-	-	+
Similarity[b]	1.5	2.0	3

[a]Composition of models A-C (+, odorant present; -, odorant absent): acetaldehyde (1.86 mg/L), 3-methylbutyl acetate (2.9 mg/L), ethyl hexanoate (0.49 mg/L), cis-rose oxide (21 µg/L), ethyl octanoate (0.63 mg/L), acetic acid (280 mg/L), (E)-ß-damascenone (0.84 µg/L), geraniol (0.22 mg/L), 4,5-dimethyl-3-hydroxy-2(5H)-furanone (5.4 µg/L), wine lactone (0.1 µg/L), ethyl isobutyrate (0.15 mg/L), ethyl butyrate (0.21 mg/L), linalool (0.17 mg/L), ethyl acetate (63.5 mg/L), 1,1-diethoxy ethane (0.37 mg/L), butan-2,3-dione (0.15 mg/L), ethyl 2-methylbutyrate (4.4 µg/L), ethyl 3-methylbutyrate (3.6 µg/L), 2-methylpropanol (52 mg/L), 3-methylbutanol (127.8 mg/L), dimethyltrisulfide (0.25 µg/L), methionol (1.41 mg/L), hexanoic acid (3.2 mg/L), 2-phenylethanol (18 mg/L), trans-ethyl cinnamate (2.0 µg/L), eugenol (5.4 µg/L), (Z)-6-dodecenoic acid-γ-lactone (0.27 µg/L) and vanillin (45 µg/L) dissolved in water/ethanol (9 + 1,w/w,1000 ml).
[b]The similarity of the model with the original Gewürztraminer wine was scored in a scale from 0 - 3: 0= none; 1= weak; 2= medium; 3= strong. Mean value of 6 assessors.

containing all odorants with OAV's ≥ 10. The aroma of model C containing all odorants with OAV's ≥ 1 was in complete agreement with that of the original Gewürztraminer wine. The latter result indicates that 29 odorants are necessary to simulate the overall flavor of Gewürztraminer wine.

Reduction of ethanol content. The significance of ethanol for the overall flavor of alcoholic beverages was already mentioned by Williams and Rosser (*23*) and Rothe and Schröder (*24*). Sensory investigations of dealcoholized Sauvignon blanc, Chardonnay Semillon and Muskat Ottonel wines were performed by Fischer et al. (*25*). The authors established that the dealcoholization process reduced the fruity attributes and the mouthfeeling of wines.

With regard to the aroma and taste of ethanol reduced wines the influence of ethanol concentrations was investigated. To take into account the taste components for the sensory evaluation these compounds were analyzed. The determination of taste components was performed according to the general procedures described in (*26*). The results are combined in Table IV. The ethanol content of model C (Table III) to which the taste compounds, detailed in Table IV, were added, was reduced stepwise and the resulting mixtures were compared with that of the original model with 100 g ethanol/L (experiment 1, Table V). The reduction of ethanol concentration to 90 g/L (experiment 2) was not noticed by the sensory panel. A further reduction to 80 g/L and 70 g/L (experiment 3) led to a weak change in the overall flavor and taste. The assessors described the mixture with more fruity and flowery odor qualities and an increasing acidic character. A reduction to 60 g/L and 50 g/L (experiment 4) led to a significant difference to the original model. The model was characterized with strong fruity and flowery notes. The mouthfeeling was diminished and a further increase of the acidic character was observed. Experiment 5 (30 g/L) and experiment 6 (10 - 0 g/L) resulted in a drastically change of the overall flavor and taste and the samples differed strongly from the original model (experiment 1). The sensory panel described the mixtures with strong fruity, flowery, acidic and adstringent aroma notes. These results indicate that ethanol reduction changes not only the flavor profile but also the taste profile.

Changes of Flavor Compounds during Wine Making of Gewürztraminer

Investigations during the different stages of wine making of Gewürztraminer wine (after pressing of grapes, after yeast fermentation, after malolactic fermentation and after maturing in high-grade steel tank) yielded a strong increase of the most potent odorants ethyl isobutyrate, ethyl butyrate, ethyl hexanoate, ethyl octanoate, 3-methylbutyl acetate, cis-rose oxide and (E)-ß-damascenone during yeast fermentation (Figure 1 and 2). After malolactic fermentation only negligible changes were recognized. During further ripening (4 months) in high-grade steel tanks an increase in concentration of wine lactone, linalool and cis-rose oxide (Figure 2), and a decrease of the amount of (E)-ß-damascenone (Figure 2), 3-methylbutyl acetate, ethyl butyrate, ethyl hexanoate and ethyl octanoate (Figure 1) was observed. In various publications (*27, 28*) the decrease of ethyl butyrate, ethyl isobutyrate, ethyl hexanoate, ethyl octanoate and 3-methylbutyl acetate during wine maturing was mentioned. The authors supposed that during aging of wines a hydrolysis of the esters occurs. The release of glycosidic bound terpene compounds, e.g. linalool, by hydrolysis and/or enzymatic reactions during wine making was reported by Williams et al. (*29*), Ayran et al. (*30*) and Gunata et al. (*31*).

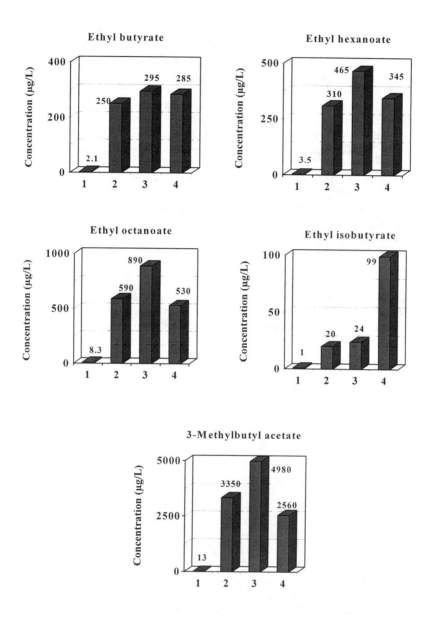

Figure 1. Concentrations of ethyl butyrate, ethyl hexanoate, ethyl octanoate, ethyl isobutyrate and 3-methylbutyl acetate in different stages of wine making; 1 = after pressing of grapes, 2 = after yeast fermentation, 3 = after malolactic fermentation, 4 = after maturing in high-grade steel tanks.

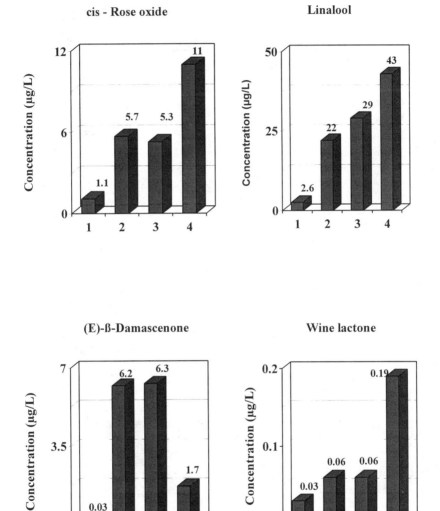

Figure 2.   Concentrations of cis-rose oxide, linalool, (E)-ß-damascenone and wine lactone in different stages of wine making; 1 = after pressing of grapes, 2 = after yeast fermentation, 3 = after malolactic fermentation, 4 = after maturing in high-grade steel tanks.

Table IV. Concentrations and taste values (≥ 0.1) of compounds in Gewürztraminer and Scheurebe wines

Compound	Concentration (mg/L)			
	Gewürztraminer		Scheurebe	
Group 1: acidic, adstringent				
Acetic acid	280	(2.3)[a]	255	(2.1)[a]
Tartaric acid	1575	(7.9)	1260	(6.3)
Citric acid	875	(2.5)	594	(1.7)
Malic acid	377	(5.0)	4790	(63)
Lactic acid	1680	(1.2)	980	(0.7)
Succinic acid	590	(12.6)	480	(10.2)
Oxalic acid	100	(2.0)	<50	(<0.1)
γ-Aminobutyric acid	21	(53)	23	(58)
Group 2: sweet				
D-Glucose	870	(<0.1)	13040	(0.8)
D-Fructose	575	(<0.1)	13500	(1.4)
Prolin	760	(0.3)	320	(0.1)
Group 3: salty				
Cl^-	20	(<0.1)	135	(0.5)
PO_4^{3-}	270	(0.4)	245	(0.3)
SO_3^{2-}	35	(0.6)	120	(2.0)
K^+	1240	(2.1)	1100	(1.9)
Ca^{2+}	32	(0.1)	231	(0.8)
Mg^{2+}	55	(0.6)	81	(0.8)
Glutamic acid	54	(0.1)	18	(<0.1)
Group 4: bitter				
Lysine	27	(0.2)	16	(0.1)

[a] Quotient of concentration in wine and taste threshold of the compound in water. Taste values were determined according to Warmke et al. (*26*).

Studies about the formation of (E)-ß-damascenone during wine making were performed by Winterhalter et al. (*32*), who identified glycosylated norisoprenoids in Riesling wine as the precursors. Laurent et al. (*33*), who investigated the influence of the malolactic fermentation on the overall flavor of Chardonnay wines, found that the concentration of butan-2,3-dione during the process increased whereas the amount of (E)-ß-damascenone had not changed. These data are in agreement with the data reported in the present study.

Table V. Influence of an ethanol reduction on the flavor and taste of
Gewürztraminer models[a]

Expt.	Ethanol Conc. (g/L)	Similarity[b]	Odor Quality	Taste
1	100	3	fruity , flowery (1.5)[c]	acidic (0.5)[c]
2	90	3	fruity, flowery (1.5)	acidic (0.5)
3	80 - 70	2	fruity, flowery (2.0)	acidic (1.0)
4	60 - 50	1.5	fruity , flowery (2.5)	acidic (2.0)
5	30	1	fruity , flowery (3.0)	acidic, adstringent (2.5)
6	10 - 0	0.5	fruity, flowery (3.0)	acidic , adstringent (3.0)

[a] The models contain the odorants and taste compounds as detailed in Table III (model C) and Table IV.
[b] The similarity of the model with the original Gewürztraminer wine was scored in a scale from 0 - 3: 0= none; 1= weak; 2= medium; 3= strong. Mean value of 6 testers.
[c] The intensity of the odor and taste quality was scored in a scale from 0 - 3: 0= none; 1= weak; 2= medium; 3= strong. Mean value of 6 assessors.

Influence of Barrel Aging on the Flavor of Gewürztraminer Wine

The differences in flavor profiles of Gewürztraminer wine aged in high grade steel tanks and Allier oak barrels, respectively, were investigated by AEDA (Table VI). In comparison to aging in high-grade steel tanks, aging in Allier oak barrels led to the appearance of 3-methylbutanal, methional, whiskey lactone, ethylguaiacol and 2,6-dimethoxyphenol and to significant higher FD-factors of vanillin, guajacol, eugenol and (E)-ß-damascenone. Quantitation experiments of the main odorants are summarized in Table VII. 3-Methylbutanal, methional, whiskey lactone, ethylguaiacol and 2,6-dimethoxyphenol were not detectable in Gewürztraminer wine aged in steel tanks. On the contrary the wines aged in Allier oak barrels contained the above mentioned compounds in a concentration range from 9.9 µg/L (methional) to 134 µg/L (whiskey lactone). Beyond that the barrel aged wine shows 2-fold higher concentrations of acetaldehyde and wine lactone, 2.5-fold higher concentration of butan-2,3-dione, 3-fold higher concentration of (E)-ß-damascenone and eugenol, 7-fold higher concentration of vanillin and 16-fold higher concentration of guaiacol than in wines aged in high grade steel tanks (Table VII). Wines aged in high grade steel tanks yielded higher amount of 3-methylbutyl acetate (2.9 mg/L) than wines aged in oak barrels (450 µg/L). The occurence of 3-methylbutanal and methional in oak barrel aged Gewürztraminer is presumably referred to oxidation reactions of the corresponding alcohols. The phenolic substances and whiskey lactone pass from the barrel into the wine (*34, 35, 36, 37*). Towey and Waterhouse (*38*) showed, that the concentrations of phenolic compounds and whiskey lactone in wines depend on the age of the oak barrel.

50

TableVI. Results of Aroma Extract Dilution Analysis of Wine Aged in Allier Oak Barrel and High Grade Steel Tanks, respectively

	FD-Factor	
Compound	Allier Oak Barrel	Steel Tank
Wine lactone	1000	1000
Ethyl 2-methylbutyrate	100	100
3-Methylbutanol	100	100
2-Phenylethanol	100	100
3-Hydroxy-4,5-dimethyl-2(5H)-furanone	100	100
Ethyl isobutyrate	10	10
Ethyl hexanoate	10	10
cis-Rose oxide	10	10
Ethyl octanoate	10	10
Linalool	10	10
Methional	10	<1
Guaiacol	10	1
Whiskey lactone	10	<1
Ethylguaiacol	10	<1
Eugenol	10	1
Vanillin	10	1
(E)-ß-Damascenone	10	1
2,6-Dimethoxyphenol	1	<1
3-Methylbutanal	1	<1
3-Methylbutyl acetate	<1	1

Calculation of OAV´s indicate that 3-methylbutanal, methional, guaiacol, ethylguaiacol and vanillin with OAV´s ≥ 1 contribute to the overall flavor of Gewürztraminer wine aged in oak barrels (Table VII), but have no significance for the wine aged in steel tanks. Furthermore the OAV of 3-methylbutyl acetate in oak barrel aged wines (OAV = 15) was by the factor of 6.5 lower than in steel tank aged wine (OAV = 97).

These results indicate that compounds with malty (3-methylbutanal), sulfurus (methional), coconut-like (whiskey lactone) and phenolic   (ethylguaiacol, guaiacol) odor qualities contribute to the overall flavor of barrel aged wines.

Literature Cited

1. Maarse, H.; Visscher, C.A. Volatile compounds in food, qualitative and quantitative data, sixth edition, TNO, Zeist: The Netherland, **1994**.
2. Acree, T.E.; Barnard, J.; Cunningham, D.G. *Food Chem.* **1984**, *14*, 273.
3. Ullrich, F.; Grosch, W. *Z. Lebensm. Unters. Forsch.* **1987**, *184*, 277.
4. Chrisholm, M.G.; Guiher, L.A.; Vonah, T.M.; Beaumont, J.L. *Am. J. Enol. Vitic.* **1994**, *45*, 201.
5. Schlich, P.; Moio, L. *Sci. Aliments* **1994**, *14*, 609.

Table VII. Influence of barrel aging on the concentrations and odor activity values (OAV´s) of potent odorants in Gewürztraminer wine

	Concentration (μg/L)			
Compound	Allier Oak Barrel		Steel Tank	
Wine lactone	0.2	(20)[a]	0.1	(10)[a]
Acetaldehyde	4320	(9)	1860	(4)
Butan-2,3-dione	405	(4)	150	(2)
3-Methylbutanal	51	(10)	<1.0	(1)
Methional	9.9	(20)	<0.5	(<1)
Guaiacol	56	(6)	3.6	(<1)
Whiskey lactone	134	-	<0.5	-
Ethylguaiacol	12	(1)	<0.1	(<1)
Eugenol	16	(3)	5.4	(1)
Vanillin	335	(2)	45	(<1)
(E)-ß-Damascenone	2.8	(56)	0.84	(17)
2,6-Dimethoxyphenol	104	(<1)	<1.0	(<1)
3-Methylbutyl acetate	450	(15)	2900	(97)

[a] Refer to footnote a in Table II.

6.   Moio, L.; Schlich, P.; Etievant, P. *Sci. Aliments.* **1994**, *14*, 601.

7.   Miranda-Lopez, R.; Libbey, L.M.; Watson, B.T.; McDaniel, M.R. *J. Food Sci.* **1992**, *57*, 985.

8.   Berger, R. G. In *Aroma Biotechnology*, Berger, R.G., Ed.; Springer-Verlag, Berlin, Heidelberg, New York, 1995, pp. 11-34.

9.   Guth, H. *J. Agric. Food Chem.* **1997**, *45*, 3022.

10.   Darriet, P.; Tominaga, T.; Demole, E.; Dubourdieu, D. *C. R. Acad. Sci. Paris* **1993**, *316*, 1332.

11.   Guth, H. In Flavour Science: Recent Developments; Taylor, A.J., Mottram, D.S., Ed.; 8th Weurman Flavour Research Symposium; The Royal Society of Chemistry: Cambridge, UK, 1996, pp. 163-167.

12.   Guth, H. *Helv. Chim. Acta* **1996**, *79*, 1559.

13.   Guth, H.; Grosch, W. *Flavour Fragrance J.* **1993**, *8*, 173.

14.   Grosch, W. *Trends Food Sci. Technol.* **1993**, *4*, 68.

15.   Rothe, M.; Thomas, B. *Z. Lebensm. Unters. Forsch.* **1963**, *119*, 302.

16.   Guth, H.; Grosch, W. *J. Am. Oil Chem. Soc.* **1993**, *70*, 513.

17.   Guth, H.; Grosch, W. *J. Agric. Food Chem.* **1994**, *42*, 2862.

18.   Allen, M.S. Lacey M.J.; Boyd S. *J. Agric. Food Chem.* **1994**, *42*, 1734.

19.   Aubry,V.; Ginies C.; Henry, R.; Etievant, P. In Flavour Science: Recent Developments; Taylor, A.J., Mottram, D.S., Ed.; 8th Weurman Flavour Research Symposium; The Royal Society of Chemistry: Cambridge, UK, 1996, pp. 331-334.

20.   Guth, H. *J. Agric. Food Chem.* **1997**, *45*, 3027.

21. Schreier, P.; Drawert, F.; Junker A. *Chem. Mikrobiol. Technol. Lebensm.* **1977**, *5*, 45-52.

22. Tominaga, T.; Masneuf, I.; Dubourdieu, D. *J. Int. Sci. Vigne Vin* **1995**,*29*,227.

23. Williams, A. A.; Rosser, P.R. *Chemical Senses* **1981**, *6*,149.

24. Rothe, M.; Schröder, R. In Flavour Science: Recent Developments; Taylor, A.J., Mottram, D.S., Ed.; 8th Weurman Flavour Research Symposium; The Royal Society of Chemistry: Cambridge, UK, 1996, pp. 348-349.

25. Fischer, U.; Berger, R.G.; Hakansson, A.; Noble, A.C. In Flavour Science: Recent Developments; Taylor, A.J., Mottram, D.S., Ed.; 8th Weurman Flavour Research Symposium; The Royal Society of Chemistry: Cambridge, UK, 1996, pp. 335-338.

26. Warmke, R.; Belitz, H.-D.; Grosch, W. *Z. Lebensm. Unters. Forsch.* **1996**, *203*, 230

27. Marais, J.; Pool, J. *Vitis* **1980**, *19*, 151.

28. Edwards, T.L.; Singleton, V.L.; Boulton, R.B. *Am. J. Enol. Vitic.* **1985**, *36*, 118.

29. Williams, P.J.; Sefton, M.A.; Wilson, B. In Flavour Chemistry: Trends and Developments, Teranishi, R. Buttery, R.G., Shahidi, F., Ed.; ACS Symp. Serie no 388, Washington, DC, USA, 1989, pp 35-48.

30. Ayran, A.P.; Wilson, B.; Strauss, C.R.; Williams, P.J. *Am. J. Enol. Vitic.* **1987**, *38*, 182.

31. Gunata, Y.Z.; Bayonove, C.L.; Baumes, R.L.; Cordonnier, R.E.; *J. Chromatogr.* **1985**, *331*, 83.

32. Winterhalter, P.; Sefton, M.A.; Williams P.J. *J. Agric. Food Chem.* **1990**, *38*, 1041.

33. Laurent, M.-H.; Henick-Kling, T.; Acree, T.E. *Weinwissenschaft* **1994**, *49*, 3.

34. Piggot, J.R.; Conner, J.M.; Melvin, J.L. *Dev. Food Sci.* **1995**, *37B*, 1695.

35. Chatonnet, P.; Dubourdieu, D.; Boidron, J.N.; Pons, M. *J. Sci. Food Agric.* **1992**,*60*, 165.

36. Rapp, A.; Versini, G. *Dt. Lebensm. Rundsch.* **1996**, *92*, 42.

37. Rapp, A. *Fresenius J. Anal. Chem.* **1990**, *337*, 777.

38. Towey, J.P.; Waterhouse, A.L. *Am. J. Enol. Vitic.* **1996**,*47*, 163.

Chapter 5

Volatile Compounds Affecting the Aroma of *Vitis vinifera* L. cv. Scheurebe

G. E. Krammer[1], M. Güntert[2], S. Lambrecht[1], H. Sommer[1], P. Werkhoff[1], J. Kaulen[1], and A. Rapp[3]

[1]Corporate Research and [2]Flavor Division, Haarmann and Reimer GmbH, P.O. Box 1253, D-37601 Holzminden, Germany
[3]Federal Centre for Breeding Research on Cultivated Plants, Institute for Grapevine Breeding Geilweilerhof, D-76833 Siebeldingen, Germany

Typical Scheurebe wines are characterized by a fruity aroma, which is described as redcurrant-like, often with a grapefruit note. In 1995 Darriet et al. identified 4-mercapto-4-methylpentan-2-one (MMP), responsible for the distinctive odor of box tree and cat urine (*1*). The perception threshold of this compound in wine has been reported to be very low (3 ng/l). Recent studies by Guth and coworkers (*2*) showed that MMP also significantly influences the aroma of Scheurebe wines. Sulfur compounds occur naturally in wines in very low concentrations, but play an important role in the determination of the flavor and aroma of the wine.
The aim of the present study was the isolation and identification of new flavor compounds from Scheurebe wine at trace level. For enrichment purposes preparative multidimensional gas chromatography (MDGC) was used. The structure elucidation of the isolated compounds was on the basis of spectroscopic methods (NMR, GC-FTIR, GC-MS) and synthesis as well. The sensory properties of the isolated compounds were correlated with the typical aroma profile of Scheurebe wines.

Numerous studies on the volatile compounds of *Vitis vinifera* wines, as reviewed by Webb (*3*), Schreier (*4*) and Rapp (*5*), helped to elucidate the basic flavor chemistry in this field of special interest. Enormous efforts were focused on the topic of varietal characterization (*6*). With regard to the analytical differentiation of the varietal aroma or "bouquet" two points of view are important. First of all it is necessary to understand the influence of specific compounds on the total flavor impression. Secondly aroma chemicals are of fundamental interest for the study of breeding experiments. A good example for this approach is the identification of 2,5-dimethyl-4-hydroxy-3(2*H*)-furanone and its methoxy derivative in berries and wines of some interspecific grapevine breedings (*7*).

EXPERIMENTAL SECTION

Wines. Two different wines *vitis vinifera* L. cv. Scheurebe from the Pfalz area in Germany were analyzed. Scheurebe is a cross breeding of the two cultivars Riesling and Silvaner. Wine no. 1: Scheurebe Qualitätswein b. A. from 1992; provided from a cooperative winery. Wine no. 2: Qualitätswein b. A. from 1995, was obtained from a small winery.

Sample preparation. For the study of trace compounds we used two different sample preparation procedures SP I and SP II for wine no. 1 and wine no. 2, respectively. The first flavor extract SP I was obtained by liquid-liquid extraction with fluorochloromethane and dichloromethane (9+1) from 45 L Scheurebe wine. For further analysis a portion of 1/3 was used. After separation on silica gel (pentane/diethyl ether) 6 fractions were analyzed. For the second flavor extract SP II we started from 200 L wine stripping off volatile compounds with vapour in a spinning cone column (SCC) system (*8*). The condensate was sequentially collected in two main portions of 8 and 2 L, respectively. The first condensate was discarded. The second condensate (2 L) was subjected to liquid-liquid extration with fluorochloromethane and dichloromethane (9+1). After separation on silica gel (pentane/diethyl ether) using medium pressure chromatography (MPLC) 4 fractions were analyzed.

Instrumental analysis. Instrumentation (capillary gas chromatography, spectroscopy) as well as analytical and preparative conditions have been described in previous publications (*9, 10*). For GC-FTIR analyses a Bio-Rad Digilab FTS-45A spectrometer connected to the Bio-Rad Tracer (Bio-Rad, Krefeld, Germany) equipped with a liquid nitrogen cooled narrow-band MCT detector and coupled to a HP 5890 series II gas chromatograph (Hewlett-Packard, Waldbronn, Germany) was applied. The samples were separated on a J&W DB-1 column (30 m x 0.25 mm; 0.25 µm film thickness) with helium as carrier gas (split injection mode). Deposition tip and transferline were held above 200°C. Absorbance spectra were recorded from 4000 to 700 cm^{-1} at a spectral resolution of 1 cm^{-1}.

For chiral separations a fused silica column (25 m x 0.25 mm, film thickness 0.25 µm) from MEGA capillary columns laboratory (Legnano, Italy) was used. The column was coated with a solution of 30% diacetyl tert. butyl silyl-β-cyclodextrin with 70% OV-1701.

Analysis of trace compounds. All fractions were checked by capillary gas chromatography (GC) with FID and sulfur specific detection (flame photometric detector, FPD; ThermoQuest CE, Egelsbach). Subsequently the different fractions were analyzed by capillary gas chromatography-mass spectrometry (GC-MS). Specific unknowns were enriched by preparative multidimensional gas chromatography (MDGC). For further structure elucidation complementary analyses using GC-MS and capillary gas chromatography-Fourier transform infrared spectroscopy (GC-FTIR) as well as ^{1}H-NMR were applied. All new compounds have been synthesized and characterized by GC-olfactometry (GC-O).

RESULTS AND DISCUSSION

Ester Compounds

2-Hydroxyglutaric acid ester. The fruity acetates are mostly synthesized enzymatically through *Saccharomyces cerevisiae* during fermentation. Esterification, as a consequence of an establishing equilibrium between acids and alcohols is further transforming the wine aroma in a significant way (*5*). The biochemical pathway of 2-hydroxyglutaric acid can be traced back to different amino acids, like e.g. glutamine, proline, arginine and histidine. After transamination and reduction the hydroxy dicarboxylic acid can be subject to esterification. The first identification of diethyl-2-hydroxy-glutarate was reported by di Stefano (*11*). Using preparative MDGC and NMR-spectroscopy we have been able to characterize ethyl-3-methylbutyl-2-hydroxy-pentanedioate (**1**) in a fraction of the liquid-liquid extract (SP I) of Scheurebe wine.

The assignment of each alcohol residue, however, was checked by synthesis starting from the lactone of 2-hydroxyglutaric acid with a two step esterification. Table 1 shows a collection of our results concerning GC-O evaluation together with retention index and mass spectral data. The sensory properties of compound **1** are described as weak, fatty, lactone-like.

1

Phenyl lactic acid esters. Furthermore we found two new esters of phenyl lactic acid. It is obvious that the formation of phenyl lactic acid starts from phenylalanine and follows a similar pathway as described above. Using preparative MDGC for the enrichment of a single peak in fraction 3 of sample preparation I (SP I), as illustrated in Figure 1, we were able to characterize two new phenyl lactic acid esters (Figure 1). The major isomer present in extract SP I was identified as 3-methylbutanol ester (**2**) with a weak, honey-like grape note in tasting. The corresponding 2-methylbutanol ester (**3**) was found in minor quantities, which leads to the question about the occurrence of the ethyl ester in wine. This compound was published by Schreier and Drawert in 1974 (*12*). Malolactic fermentation is the main source for lactate derivatives in wines. Therefore it is assumed that phenyl lactic acid esters are the result of esterification after the end of fermentation.

2   **3**

Table I: MS-data and sensory description of some new flavor compounds in two different Scheurebe extracts (SP I, SP II)

No.	Compound name	Sensory description[a]	Extract	RI (DB-1)	MS-data (m/z, %)
1	ethyl-3-methylbutyl-2-hydroxy-pentanedioate	weak, fatty, lactone-like	SP I[b]	1595	173 (5), 159 (10), 131 (7), 103 (9), 85 (100), 71 (15), 55 (19), 43 (22), 42 (13), 41 (13)
2	phenyl lactic acid, 3-methylbutyl ester	weak, honey-like, grape-note	SP I	1654	218 (6), 149 (26), 148 (32), 147 (10), 131 (11), 121 (40), 103 (34), 92 (25), 91 (100), 77 (16), 43 (31), 41 (18)
3	phenyl lactic acid, 2-methylbutyl ester	weak, woody, dusty	SP I	1658	218 (5), 149 (19), 148 (53), 147 (14), 131 (14), 121 (45), 103 (36), 92 (27), 91 (100), 77 (18), 65 (14), 43 (31), 41 (18)
4	thiocarbonic acid O, S-dimethylester	fruity, overripe, sulfur note	SP II[c]	761	**106** (80), 75 (49), 61 (18), 59 (60), 47 (100), 46 (34), 45 (61), 29 (20)
5	succinic acid ethyl methionyl ester	methionol-like, sulfur note	SP I	1641	**234** (1), 129 (10), 101 (52), 89 (43), 88 (100), 73 (53), 61 (30), 55 (17), 41 (23)
6	methionol-S-oxide	overripe, cheesy, putrid	SP I	1185	**122** (16), 105 (15), 104 (30), 78 (59), 76 (36), 64 (96), 61 (26), 59 (48), 57 (34), 47 (49), 41 (79), 31 (100)
10	2-methyl-1,3-oxathiane	chemical, burnt note	SP II	907	**118** (45), 103 (45), 74 (72), 59 (15), 47 (16), 46 (99), 45 (42), 43 (31), 27 (17)

[a] Sensory description by GC-O
[b] sample preparation I: liquid-liquid extraction
[c] sample preparation II: spinning cone column

Volatile Sulfur Compounds

The origin of sulfur compounds in wine. In plants the basic source of sulfur is sulfate. After biochemical reduction numerous sulfur containing molecules are formed. The most abundant S-compounds, however, are represented by two amino acids, methionine and cysteine. The crushing of the grapes initiates further enzymatic transformation arising from grape juice enzymes. Later during the fermentation step yeasts also are able to reduce sulfate for their own metabolism, which is in fact another introduction of sulfur into wine chemistry (*13*). It is known that early stages of fermentation are characterized by the growth of non-Saccharomyces species, mostly apiculate yeasts from *Kloeckera* and *Hanseniaspora* genera. Gil and coworkers found, that the initial presence of apiculate yeasts leads to production of some compounds, like e. g. higher alcohols (*14*). Little, however, is known about the influence on sulfur compounds.

Beside the conversion of endogenic sulfur compounds the addition of S-compounds like sulfite, as an antimicrobial agent, antioxidant and enzyme inhibitor (*15*) or like thiamine (vitamin B1), as a nutrient for yeasts, are allowed in the EEC (within defined maximum values). Furthermore chemical reactions like sulfite addition to aldehydes, Maillard reaction or Strecker degradation play an important role with regard to the sulfur chemistry of wines.

Thiocarbonic acid derivative. Carbonic acid is a major product from fermentation. It is liberated by enzyme mediated pyruvate decarboxylation. Carbonylsulfide (COS), a formal derivative of carbonic acid has been found in a variety of wines at concentrations in the 10 μgL^{-1} range (*15*). The identification of thiocarbonic acid O, S-dimethyl ester (**4**) in Scheurebe wine extract (SP II), however, is a new finding (Figure 2). We checked the spectroscopic data by synthesis starting from methyl chloroformate and sodium thiomethoxide. Despite the fact that compound **4** is known for a long time in literature, the occurrence in nature has not been published until now. Considering the unusual structure of this molecule it is evident that the formation does not follow known biochemical pathways. According to Vreeken et al. compound **4** was observed as a reaction product between carbon disulfide and methanol on active charcoal (*16*). Methanol and CS_2 as well are constituents of wines. The investigation of model systems, however, is beyond the scope of this study. At present it has to remain unclear, whether this compound is of endogenous or anthropogenic origin. We checked the sensory properties of carbonate **4** by GC-O and tasting. Different amounts of **4** were added to Silvaner wine samples. The odor was characterized by an overripe, fruity, sulfur note, which dominated the aroma of the wine samples down to a concentration of 0,1 μgL^{-1}.

4

Methionol derivatives. Methionol, 3-(methylthio)-1-propanol, a well known S-compound in white and red wines was first identified by Muller et al. (*17*). The biosynthesis of methionol starts from methionine following the Ehrlich-mechanism (*18*). The sensory properties of methionol are descibed as raw potatoes (*17*). Not only methionol, but also succinic acid is a normal by-product of alcoholic fermentation. Our studies on volatile compounds in Scheurebe wines revealed for the first time the presence of succinic acid ethyl methionyl ester (**5**). This compound is characterized

58

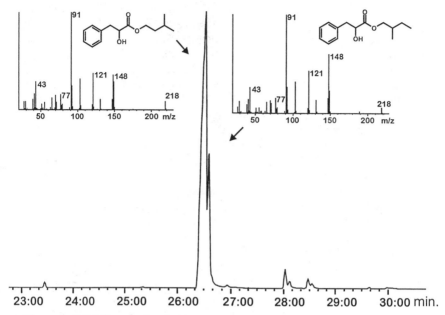

Figure 1. GC-MS analysis of phenyl lactic acid 3-methylbutyl ester (**2**) and of the corresponding 2-methylbutyl ester (**3**) after sample preparation I (SP I) and subsequent preparative enrichment using MDGC. GC column: DB-1.

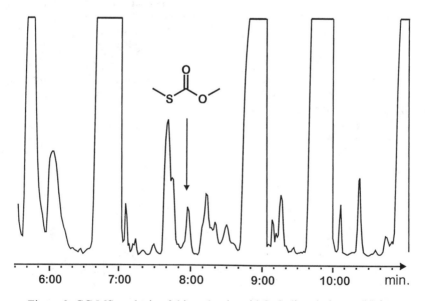

Figure 2. GC-MS analysis of thiocarbonic acid O, S-dimethyl ester (**4**) in Scheurebe wine extract (SP II). GC column: DB-Wax

by a similar taste like methionol itself (Table I). The flavor perception, however, occurs delayed in comparison to the free alcohol. According to the work of Rapp et al. (19) bottle maturation of wine shows significant changes in ester contents. In particular the increase of dicarboxylic acid ethyl esters is strongly influenced by storage time, temperature and pH. Regarding these results it seems to be possible that the formation of succinic acid ethyl methionyl ester is depending on individual storage conditions.

5

Our investigations starting from an extract obtained by spinning cone countercurrent extraction (SP II) led us to another new methionol derivative, methionol-S-oxide (**6**). The presence of this compound in wine emphasizes the pervasive influence of oxygen during wine making. Although the formation of sulfoxides in plants is catalyzed by high specific enzymes, as it is known in *Allium* species, methionol-S-oxide probably originates from chemical oxidation. Figure 3a shows the GC-FTIR spectrum of **6**, which is dominated by strong stretching vibrations for O-H at 3320 cm^{-1} and for S=O at 1058 cm^{-1}. The enantiodifferentiation of compound **6** was achieved on a chiral GC column coated with diacetyl tert. butyl silyl-β-cyclodextrin (Figure 3b). Parallel to our work Küsters and coworkers recently published similar results on the separation of racemic aryl and methylthio sulfoxides (20). This application clearly elucidates the presence of a pyramidal structure in sulfoxides as result of an free electron pair at the sulfur atom.

6

Cyclic sulfur compounds. As mentioned in the paragraphs before, the yeast metabolism is responsible for the biogeneration of important flavor compounds. Holding wines in contact with yeast lees (sur lie) for extended time after fermentation is an additional source for specific flavor notes. The GC-MS analysis of fraction 5 obtained after preseparation of the liquid-liquid extract on silica gel (SP I) and further enrichment using preparative MDGC revealed the presence of 2-methyl-3-furanthiol acetate (**7**). The free thiol was characterized in yeast extracts before (21, 22). Most recently, Etiévant expressed his opinion that the disulfide of 2-methyl-3-furanthiol, *bis*-(2-methyl-3-furyl)disulfide, can occur in some white wines stored on lees such as Meursault, Puligny Montrachet, Chassagne Montrachet, Muscadet, Sherry and Champagne (23). The degradation studies of thiamine, conducted by Güntert and coworkers (24) showed, that many heterocycles bear the basic structure of 2-methyl-3-furanthiol. Therefore it seems to be evident that thiamine degradation has an

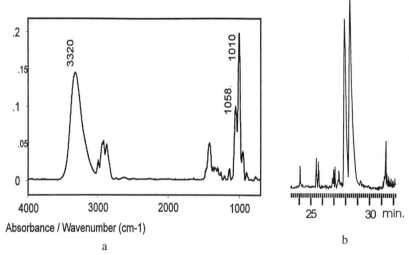

Figure 3a. GC-FTIR spectrum of methionol-S-oxide (**6**) with stretching vibrations for O-H at 3320 cm^{-1} and for S=O at 1058 cm^{-1}.
Figure 3b. Enantiodifferentiation of methionol-S-oxide (**6**) on a chiral GC column coated with diacetyl tert. butyl silyl-β-cyclodextrin.

influence on the formation of compound **7**. In Scheurebe wine we found furan **7** to be present at trace level (Figure 4, top).

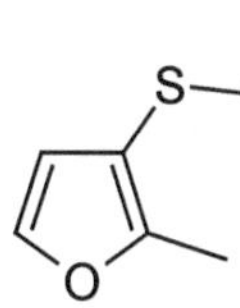

7

Using SCC extraction (SP II) we obtained an authentic extract of Scheurebe wine, which was subject to detailed GC-MS analysis. In this extract we were able to identify 2-formyl-thiophene (**8**) for the first time (Figure 4, middle). Compound **8** is known to be formed during thermal degradation of thiamine in the presence of cysteine (*24*). Similar results were obtained by Silwar et al. (*25*) using a cysteine/methionine/furfural model system. Shibamoto showed that the reaction of hydrogen sulfide together with furfural can also lead to thiophene **8** (*26*). Recently the same group reported about the reaction of short chain aldehydes with hydrogen sulfide from amino acids to form thiophenes (*27*). During fermentation hydrogen sulfide is a product of the sulfate reduction pathway and therefore a possible key intermediate in the formation of thiophene **8**. Well known thiophene derivatives in wines are cis- and trans-tetrahydro-2-methylthiophene-3-ol (*28*) and the corresponding product from oxidation, 2-methyltetrahydro-thiophene-3-one (*29*). Other thiophene derivatives are 2,4-di-tert.-butylthiophene (*30*) as well as 2-(1-butyl)-5-(2-methyl-propyl)-thiophene and 2,5-di-2-methylpropyl-thiophene (*31*). The sensory evaluation of S-compound **8** revealed roasty, coffee-like notes.

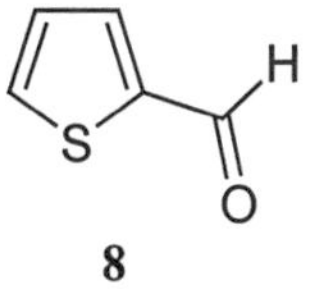

8

These sensory properties are also important for the odor of furfurylmethylsulfide (**9**). Again the countercurrent extraction (SCC II) proved to be a good approach to the analysis of this compound in wine (Figure 4, bottom). Previously different groups reported the presence of furfuryl derivative **9** in roasted coffee beans (*33, 34*), which indicates the influence of Maillard type reactions during wine making.

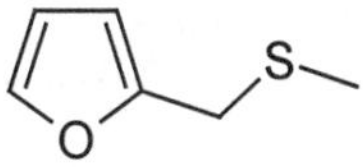

9

In the same extract (SP II) we found another new compound in wine flavor, 2-methyl-1,3-oxathiane (**10**). The formation of this heterocyclic sulfur compound with chemical, burnt odor properties can be explained as a mixed acetal of 3-mercaptopropanol with acetaldehyde. 3-Mercaptopropanol, however, is not known as

62

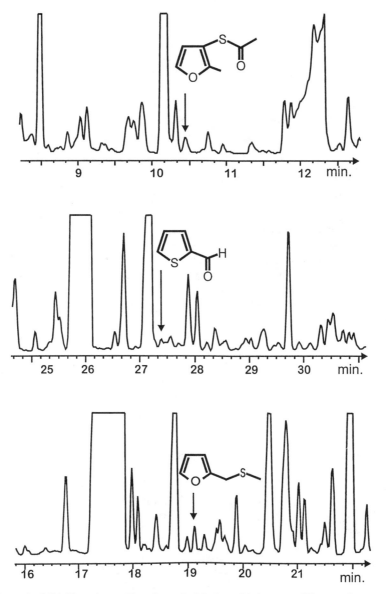

Figure 4. GC-MS analyses. Top: 2-methyl-3-furanthiol acetate (**7**), sample preparation I (SP I), GC column: DB-1. Middle: 2-formyl-thiophene (**8**), SP II, GC column: DB-Wax. Bottom: furfurylmethylsulfide (**9**), SPII, GC column: DB-Wax.

a wine constituent. On the other hand the demethylation product of methionol can be regarded as a direct degradation product of thiamine (*24*). At the present state of this study, however, the origin of oxathiane **10** has to remain unclear.

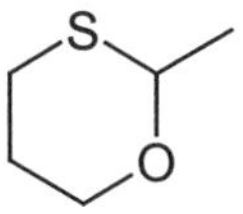

10

The most significant correlation to the influence of yeast material is the identification of 4-methyl-5-vinyl-thiazol (**11**) in fraction 3 the liquid-liquid extract (SP I). This compound is a dehydration product of sulfurol, which is a primary degradation product of thiamine with roasty, fatty, peanut-like flavor notes (*24*).

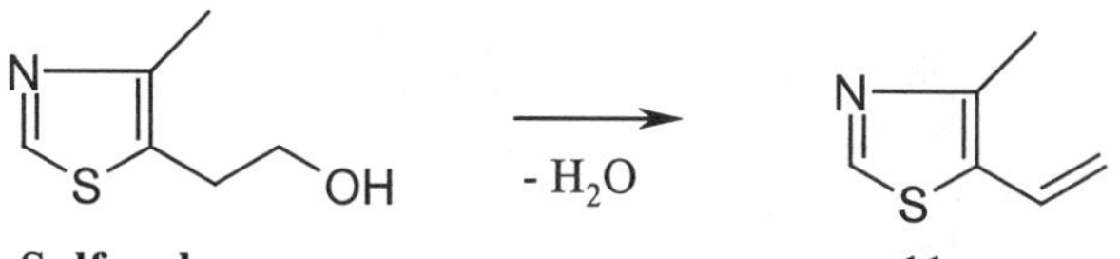

Sulfurol $- H_2O$ **11**

Flavor compounds in ppt-level

On the basis of GC-O analyses together with retention index evaluation we were able to recognize a flavor compound with sensory properties similar to 4-mercapto-4-methylpentan-2-one (MMP) in fraction 2 of the SCC-extract (SP II). The same observation was already reported by Rapp and Pretorius in 1990 (*34*). According to Guth MMP has to be regarded as a keycompound for the typical blackcurrant-note in the flavor of Scheurebe (*2, 35*). Further importance to Scheurebe flavor is attributed to the 3S, 3aS, 7aR-isomer of wine lactone (3a, 4, 5, 7a-tetrahydro-3,6-dimethyl-2(*3H*)-benzofuranon) by this author. In our study other important aroma chemicals like e.g. (E)-*β*-damascenone were found in both extracts (SP I + SP II) in significant amounts. In order to understand the sensory relevance of all mentioned trace compounds, however, further studies on model systems are necessary.

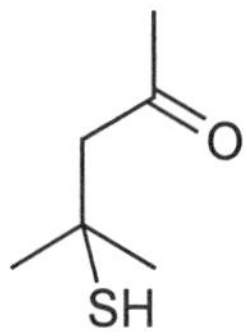

MMP

Conclusions

The trace analysis of flavor compounds in two Scheurebe wines revealed the presence of 11 new molecules. On one hand the structure of thiocarbonic acid, O-, S-dimethyl ester is unusual according to the known biochemical pathways in wine chemistry. On the other hand the identified esters of phenyl lactic acid, 2-hydroxy glutaric acid and succinic acid corroborate the knowledge about ester formation during fermentation and storage. In addition methionol-S-oxide, as an oxidation product of methionol, illustrates the pervasive influence of oxygen during wine making. Using a chiral GC column the enantiodifferentiation of methionol-S-oxide was demonstrated.

For the first time the identication of 2-methyl-3-furanthiol acetate, furfuryl-methylsulfide, 2-formyl-thiophene and 4-methyl-5-vinyl-thiazol confirms the assumption that products from thiamine degradation and yeast lees are present in wines at trace level.

Acknowledgements

We would like to thank the management of Haarmann & Reimer for permission to publish this paper, the staff of the H&R research department for their valuable and skillful work. We are indebted to Dr. Ian Gatfield for helpful discussions. Furthermore we are grateful to Dr. Mothes for SCC extraction and to Mr. G. Kindel for sensory evaluations.

Literature Cited

1. Darriet, P.; Tominaga, T.; Lavigne, V.; Boidron, J.-N.; Dubourdieu, D. *Flavour Fragrance J.*, **1995**, *10*, 385-392.
2. Guth, H. *Lebensmittelchemie*, **1997**, *51*, 4.
3. Webb, A.D.; Muller, C. J. In *Advances in Applied Microbiology*; Perlman, D. Ed., Academic Press, New York, **1972**, pp 75-146.
4. Schreier P. *CRC Food Sci. Nut.*, **1979**, *12*, 59-111.
5. Rapp, A.; Mandery, H. *Experentia*, **1986**, *42*, 873-884.
6. Rapp, A.; Versini, G. In *Food Flavors: Generation, Analysis and Process Influence*; Charalambous, G., Ed., Elsevier Science B.V., Amsterdam, **1995**, pp1659-1694.
7. Rapp, A.; Knipser, W.; Engel, L.; Ullemeyer, H.; Heimann, W. *Vitis*, **1980**, *19*, 13-23.
8. Flavourtech, Spinning Cone Column product information, **1996**.
9. Güntert, M; Brüning, J.; Emberger, R.; Köpsel, M.; Kuhn, W.; Tielmann, T.; Werkhoff, P. *J. Agric. Food Chem.*, **1990**, *38*, 2027.
10. Güntert, M.; Brüning, J.; Emberger, R.; Hopp, R.; Köpsel, M.; Surburg, H.; Werkhoff, P. In *Flavor Precursors - Thermal and Enzymatic Conversions*; Teranishi, R., Takeoka, G. R., Güntert, M., Eds.; ACS Symposium Series 490; American Chemical Society, Washington DC, **1992**, pp. 140.
11. Stefano di R. *Vini d'Italia*, **1981**, *23*, 29.

12. Schreier, P.; Drawert, F. *Chem. Mikrobiol. Technol. Lebensm.*, **1974**, *3*, 154.
13. Rauhut, D. In *Wine Microbiology and Biotechnology*; Fleet, G. H., Ed. Gordon and Breach Science Publishers, **1993**, 183-223.
14. Gil, J. V.; Mateo, J. J.; Jiménez, M.; Pastor, A.; Huerta, T. *J. Food Science*, **1996**, *61*, 1247-1249.
15. Shooter, D. *Chemistry in New Zealand*, **1994**, May, 30-34.
16. Vreeken, R. J.; Niessen, W. M. A.; Van Thuijl, J. *Int. J. Environ. Anal. Chem.*, **1988**, *33*, 23-33.
17. Muller, C. J.; Kepner, R. E.; Webb, A. D. *Am. J. of Enol. Vitic.*, **1971**, *22*, 156-160.
18. Bärwald, G.; Kliem, D. *Chemie Mikrobiologie Technologie der Lebensmittel*, **1971**, *1*, 27-32.
19. Rapp, A.; Marais, J. In *Shelf Life Studies of Foods and Beverages - Chemical, Biological, Physical and Nutritional Aspects*; Charalambous, G., Ed.; Elsevier Science Publishers, Amsterdam, **1993**; 891-921.
20. Küsters, E.; Gerber, G. *Chromatographia*, **1997**, *44*, 91-96.
21. Ames, J. M.; MacLeod, G. *J. Food Sci.*, **1985**, *50*, 125.
22. MacLeod, G.; Ames, J. M. *Dev. Food Sci.*, **1986**, *13*, 263.
23. Etiévant, P. In *Volatile Compounds in Foods and Beverages*; Maarse H., Ed.; Marcel Dekker, **1991**, 525.
24. Güntert, M; Bertram, H.-J.; Hopp, R.; Silberzahn, W.; Sommer, H.; Werkhoff, P. In *Recent Developments in Flavor and Fragrance Chemistry*; Hopp, R.; Mori, K., Eds.; VCH Publishers, Weinheim, **1993**, 215-240.
25. Silwar, R.; Tressl, R. Z. *Lebensm.-Unters. Forsch.*, **1989**, *189*, 205.
26. Shibamoto, T. *J. Agric. Food Chem.*, **1977**, *25*, 206-208.
27. Shibamoto, T. In *Contribution of Low- and Non-Volatile Materials to the Flavor of Foods*; Pickenhagen, W., Ed.; Allured Publishing Corporation, Carol Stream, **1996**, pp 183-192.
28. Rapp, A.; Güntert, M.; Almy, J. *Vitis*, **1984**, *23*, 66-72.
29. Schreier, P.; Drawert, F.; Junker, A.; Barton, H.; Leupold, G. Z. *Lebensmittel-Unters.-Forsch.*, **1976**, *162*, 279-284.
30. Shimizu, J.; Watanabe, M. *Agric. Biol. Chem.*, **1982**, *46*, 1377-1380.
31. Shimizu, J.; Watanabe, M. *Agric. Biol. Chem.*, **1982**, *46*, 2353-2355.
32. Gianturco, M. A.; Giammatino, A. S.; Friedel, P.; Flanagan, V. **1964**, *20*, 2951-2958.
33. Gutmann, W.; Werkhoff, P.; Barthels, M.; Vitzthum, O. G. *Colloq. Sci. Int. Cafe, [C. R.]*, **1977**, *8*, 153-161.
34. Rapp, A.; Pretorius, P. J. In: *Flavours and Off-Flavours '89*; Charalambous, G. Ed.; Elsevier Amsterdam, New York, **1990**, pp. 1-21.
35. Guth, H. In *Flavour Science - Recent Developments*; Taylor, A. J., Mottram, D. S. Eds.; The Royal Society of Chemistry, **1996**, pp 163-167.

Chapter 6

Yeast Strain and Wine Flavor: Nature or Nurture?

J. H. Thorngate, III

Department of Food Science and Toxicology, University of Idaho,
Moscow, ID 83844-1053

The causal effects of *Saccharomyces cerevisiae* on wine flavor production are well documented. The causal effects of distinct *S. cerevisiae* strains differentially affecting wine flavor are less well demonstrated. While different strains have been found to differentially affect both volatile and macromolecule composition in actual wine and model solutions, these differences have not been unambiguously shown to carry over to the human perceptual space. Nevertheless, popular opinion regarding yeast strain effects has reached mythic status despite the lack of unequivocal supporting evidence. Recent research has once again focused on direct comparisons of different *S. cerevisiae* strains; it is suggested that the real task should be to determine the intra-strain versus inter-strain sensory variability and the dependence of the intra-strain variability on extrinsic factors.

Joseph Campbell has examined the varying bases of myths, whether deistic, cosmological or sociological. These myths serve four primary functions according to Campbell (*1*), with the sociological myths "supporting and validating a certain social order." The wine community, a distinct culture unto itself, has never been particularly immune to mythology; various aspects of ritual in the American wine culture have been discussed by Fuller (*2*), whom states that:

> The creeds of wine culture are numerous. They pertain first and foremost to beliefs about what makes any given wine truly excellent…Sectarian opinions exist, of course, but there is nonetheless a kind of inherited orthodoxy about the respective roles of climate, aging, relative proportion of residual sugar to acidity, etc.

The danger of these creeds is that they become rapidly subverted into dogma, which, as Martini and Martini (*3*) note, impedes scientific progress.

A variety of common myths encountered in winemaking concern the yeasts used in vinification. The unquestioning acceptance and prevalence of these myths is demonstrated from the following postings, taken from the Internet usenet group *alt.food.wine*:

```
Subject:      Yeasts -- Wild (Natural) vs. Cultured
Date:         1997/02/11

In the areas of France which consistently rely on wild yeasts,
vineyards have been planted there "forever." Wild yeast was used
because Louis Pasteur hadn't discovered yeast, and no one knew what
is was or how to culture it.  In a monocultural environment (such as
Bordeaux) - where nothing is really growing except grapes, and has
been that way for
centuries -- the so-called wild yeasts have evolved over the
generations into a relatively pure strain.

California has not had vineyards planted for centuries.  Most
vineyards have been planted since the 1960s, and in many cases, in
areas which have never before grew grapes.  Some areas, such as
parts of Napa Valley, are becoming monocultural environments; other
areas, such as Monterey or Santa Barbara, are not.  In these latter
areas, other crops also grow in and around the vineyards, and the
naturally-occurring yeasts will *not* have evolved to such a pure
strain, and problems *can* result in their use.  Eventually
monocultural environments should yield a "wild" yeast pure enough to
use without problems.  It is only a matter of evolution, and that
takes time.



Subject:      Re: Yeasts -- Wild (Natural) vs. Cultured
Date:         1997/02/23

There are certainly flavor differences imparted and/or accentuated
by yeasts.  Of course, if the yeasts in a given area/vineyard/cellar
don't give good wine, there may be little alternative but to use
cultured yeasts.  What seems clear is that in most cases there are
textural and aromatic differences between wines from the same
vineyards made from indigenous yeasts and from cultured yeasts.  The
generalization runs along these lines:  the c.y. wines tend toward
brightness of the particular fruit the strain emphasizes and their
textures are generally more brash, more forceful.  The i.y. wines
tend to have lusher textures and less imposed simplicity of flavors.
This is, of course, one of those (ill-defined) all-things-being-
equal situations.
```

This thread highlights two common myths regarding the yeasts involved in winemaking:

1) that yeasts indigenous to the vineyard have been naturally selected for over time to pure strains which optimize wine quality (and are thus preferable to cultured yeasts)
2) that yeast strains are unique in their contributions to the perceptual sensory properties of wine

Whether or not the prevalence of these myths has to do more with neo-Luddite attitudes towards modern biotechnology or more with a bias towards Old World "wisdom" is unclear. However, ample evidence collected over the past forty years has demonstrated that the first myth is untenable (4), and the second of limited, if not questionable, import.

Ecology of *S. cerevisiae*

That the berry bloom is one source of fermentation inoculum is indisputable; yeast localize near the pedicels and stomata of intact grapes, and near sites of epidermal injury (5). These yeasts are, however, predominately apiculate (*i.e.*, *Kloeckera apiculata*, the imperfect stage of *Hanseniaspora uvarum*); *Saccharomyces* species are only infrequently found on grape surfaces (6). Indeed, in a sampling of 810 grapes collected during the 1980 vintage, Rosini and coworkers isolated only one species of *Saccharomyces*, *S. bayanus*; no *Saccharomyces* species were isolated during the previous vintage (7). More recent research isolating yeasts from grape surfaces found that, when *S. cerevisiae* species were present, they never exceeded 10 colony forming units/cm^2 (8).

Given that *S. cerevisiae* is the dominant yeast species at the end of fermentation (9) the question then becomes does *S. cerevisiae* dramatically proliferate from low initial cell counts, or are there alternative sources of *S. cerevisiae* inoculum? Martini and Martini (3) contend that the latter represents the true situation, as *Saccharomyces* species are not commonly indigenous to the surfaces of wild species of fruits and berries (10), but rather rapidly colonize the winery and winery equipment. This was elegantly demonstrated by Rosini (11) using the hydrogen sulfide negative strain DBVPG 1739. Following two years of use as a starter culture, the H$_2$S$^-$ strain had colonized the winery; DBVPG 1739 dominated an uninoculated fermentation in the third year, overwhelming the grape-indigenous *Kl. apiculata* strains. Similar results were found by Constantí *et al.* (12), whom found that *S. cerevisiae* MF01 rapidly colonized a winery over a period of two years after being used as the sole inoculum in the first year. However, two indigenous strains also were isolated in the second year, indicative of a vineyard contribution. Frezier and Dubourdieu (13) also found that consecutive vintages of spontaneously fermented wine were dominated by one strain of *S. cerevisiae*, although the authors could not demonstrate the point of origin for the yeasts.

Certainly not all research supports the theory that *S. cerevisiae* inoculum is a winery-driven process, however. Schütz and Gafner (14), in a study of spontaneous fermentation of Muller-Thurgau and Pinot noir grapes over two vintages, found that the yeast strain populations varied both by must and vintage. This would not be expected if winery-indigenous strains served as the prime inoculums. Fleet *et al.* (15) also found must-specific strains over one vintage, but their results were confounded by having conducted the fermentations in different facilities. Querol *et al.* (16), using mitochondrial DNA restriction endonuclease analysis, also found must-specific strain differences, but again the results were confounded by fermentation facility.

Longo *et al.* (*17*) did not find evidence for either climate or vintage affecting the occurrence of *S. cerevisiae*, nor did Regueiro *et al.* (*18*) find evidence for geographical effects. However, neither of these research groups karyotyped the yeasts to determine specific strain variations. In contrast, Mortimer (*19*) has reported findings from more extensive genetic studies in which different fermentations conducted at the same wineries expressed different yeast populations. As Mortimer (*19*) notes, this supports a vineyard as opposed to a cellar source for the inoculum, in agreement with the findings of Schütz and Gafner (*14*). The research of Vezinhet *et al.* (*20*) concurs with this conclusion; utilizing molecular identification techniques they found evidence for locale-specific strains.

While Vaughan-Martini and Martini (*4*) consider the vineyard origin a "myth," and Mortimer (*19*) considers the vineyard the true point of origin, the most likely explanation would appear to fall between the two extremes. Thus domination of a fermentation by the grape microflora vs. the indigenous wine microflora would depend upon such factors as climatic conditions, fungicide use, grape variety, vinification practises, and winery sanitation regimens (*21*). Regardless, in neither the vineyard nor the winery are the conditions appropriate to naturally select for a pure strain of yeast. As Oliver (*22*) notes, strain development/selection is a chancy proposition at best, given that yeasts are frequently unbalanced polyploids, with multiple genes controlling any specific trait. Kunkee and Bisson (*23*) likewise caution against presupposing that the genetic make-up of any culture remains stable; mounting evidence suggests rather that the inherent genetic instability may cause significant phenotypic changes. Indeed, *Mortimer et al.* (*24*) proposed that yeast regularly undergo a process they termed "genome renewal," in which heterozygous recessive mutants transform into homozygous diploids. This is supported by Vezinhet *et al.*'s (*20*) finding of nuclear and mitochondrial DNA polymorphism in indigenous *S. cerevisiae* strains.

Finally, as Martini and Martini (*3*) note, the selective pressures at work in the winery environment may influence the survivability of the indigenous yeasts with respect to ethanol or sulfur dioxide tolerance; it is not clear, however, how such environmental pressures would be selecting for desired flavor production. Certainly few winemakers completely sterilize their wineries and scorch their vineyards when they obtain a wine with less than optimal flavor attributes!

S. cerevisiae and Wine Flavor

Even though there is no evidence for natural selection of yeasts in either the vineyard or the winery for optimizing flavor attributes, this does not preclude the possibility that yeast strains are specific in their effects on the sensory properties of wine. A distinction must be made, however, between observable chemical differences and practical sensory perceptual differences, as measurable chemical differences may be imperceptible or indistinguishable to the human observer (*25*). It is also not uncommon that immeasurable chemical differences are of significant perceptual importance, further complicating attempts at drawing causal chemical-sensory relationships (*5*).

Chemical Differences. Wine aroma is comprised of on the order of 600-800 distinct volatile compounds, with a total concentration of approximately 1.0 g/L (*26*). Of these compounds, some originate with the grapes, some from the fermentation process, and the remainder from such processing factors as oak exposure and length of bottle aging (*27*). As the fermentation process is the primary source of wine aroma (*28*), it is seems logical enough to assume that the yeast strain used has a significant effect on volatile production.

Berry and Watson (*29*) have classified yeast aroma production into five general chemical categories: alcohols, esters, carbonyl compounds (aldehydes and ketones), sulfur-containing compounds and organic acids. Of these, the carbonyl compounds and organic acids tend to have minimal sensory impact (*27*), with the notable exceptions of acetaldehyde, diacetyl and acetic acid; the higher alcohols (fusel alcohols), esters and sulfur-containing compounds, however, contribute significantly to yeast-derived aroma.

With regards to carbonyl compounds, acetaldehyde is the predominant aldehyde formed during fermentation; most aldehydes produced, however, are formed independent of direct yeast action (*29,30*). Delteil and Jarry (*31*) found significant strain-specific production differences–*S. cerevisiae* strain K1 producing 128 mg/L and *S. cerevisiae* strain D47 producing 105 mg/L; Ough and Amerine (*32*) report average acetaldehyde concentrations in wines on the order of 54 mg/L.

Of the keto compounds, Soufleros and Bertrand (*33*) reported that diacetyl (2,3-butanedione) concentrations varied considerably among the fifty yeast strains, with concentrations ranging from 16 to 1373 mg/L. As typical concentrations in wine average less than 2 mg/L (*32*) it is assumed that the values were actually in µg/L; the authors provided no alternative explanation for the reported values. Martineau *et al.* (*34*) recently analyzed forty-one United States' Chardonnays and found diacetyl concentrations ranging from 0.0005 to 1.7 mg/L, with an average concentration of 0.38 mg/L.

Acetic acid represents the only organic acid of normal olfactory sensory significance to wines (*27*). While acetic acid may be formed by *S. cerevisiae*, the concentrations produced are typically less than 300 mg/L (*35*), far lower than the concentrations produced by spoilage microorganisms (*i.e.*, *Acetobacter*) which are the predominate source (*32*). Apiculate yeasts may also serve as a source of acetic acid; Romano *et al.* (*36*) found that *Kl. apiculata* typically produced greater than 200 mg/L acetic acid in synthetic medium fermentations.

Fusel alcohols (1-propanol, 2-methyl-1-propanol, 2-methyl-1-butanol, 3-methyl-1-butanol, 1-hexanol, 2-phenyl-ethanol) were actually among the first aroma constituents studied, as early gas chromatographic research had indicated, erroneously, that these compounds represented the predominant volatile fraction in wines (*28*). Yeast-specific fusel alcohol production has been studied by a number of researchers (*31,33,37-39*), all of whom found production differences among yeast strains. Unfortunately yeast strains have not usually been replicated among studies; an exception is the work of Delteil and Jarry (*31*) and Kunkee and Vilas (*39*). Their results for the fusel alcohol isobutanol (2-methyl-1-propanol) are shown in Table I. Soufleros and Bertrand (*33*) studied fifty different yeast strains; unfortunately their data do not allow for statistical analysis. Mateo and coworkers (*38*) examined ten

(*39*). Their results for the fusel alcohol isobutanol (2-methyl-1-propanol) are shown in Table I. Soufleros and Bertrand (*33*) studied fifty different yeast strains; unfortunately their data do not allow for statistical analysis. Mateo and coworkers (*38*) examined ten strains, again without statistical analysis, although it is clear from their data that significant differences existed among strains; *e.g.*, *S. cerevisiae* var. *cerevisiae* produced 56.2 mg/L of isoamyl alcohol, whereas *S. cerevisiae* var. *chevalieri* produced 153.0 mg/L.

Table I. Isobutanol Production (mg/L) for *S. cerevisiae* Strains D47 and K1

Source	Strain D47	Strain K1	L.S.D.[a]
31	18.4	17.12	1.02
39	21.1	19.92	1.56

[a]Fisher's Least Significant Difference, 5%

Esters, not fusel alcohols, actually comprise the most abundant group of volatile compounds in wines; Rapp (*26*) has listed over 300 esters and lactones found in grapes, musts and wines. The esters are largely responsible for the fruity aromas associated with wine (*32*), especially young wine (*27*). Of the esters, ethyl acetate predominates by some two orders of magnitude (see *40*); however, the low aroma thresholds of a number of the fatty acid ethyl esters makes them of sensory import nonetheless (*27*).

Differences in ester production have been consistently found among yeast strains (*31,33,37-39,41-43*). As was noted above the lack of inter-study replication lessens the ability to ascertain validity; furthermore, the values reported for individual esters can vary widely among studies. Results for isoamyl acetate are presented in Table II.

Table II. Isoamyl Acetate Production by Various Yeast Strains

Study	Yeast Strain	N^a	[Isoamyl Ac.][b]
41	*S. cerevisiae*	2	0.4 - 8.4
	P. fermentans	1	0.5 - 9.3
33	*S. cerevisiae*	12	1.2 - 3.5
	S. bayanus	3	0.9 - 1.6
42	*S. cerevisiae*	6	1.7 - 2.9
	S. bayanus	3	1.7 - 1.9
43	*S. cerevisiae*	17	0.6 - 14
37	*S. cerevisiae*	2	2.9 - 7.0
31	*S. cerevisiae*	2	9.0 - 16

[a]Number of strains
[b]Isoamyl acetate concentration, mg/L

The volatile sulfur-containing compounds occurring in wines, while few in number (Rapp has listed 20; *26*), can be especially problematic for wine quality. As the sensory thresholds for these compounds tend to be quite low, on the order of μg/L (44), slight variations in concentration can have major sensory repercussions. While the majority of these compounds are not synthesized by yeasts (29), Rankine (45,46) found strain-specific effects in the ability to produce hydrogen sulfide. Eschenbruch and coworkers (47), however, found no significant differences in hydrogen sulfide production among the twelve yeast strains they studied. And Zeeman et al. (43), while finding strain-specific differences, concluded that the absolute amount of hydrogen sulfide produced may be of less significance than the stage of fermentation in which the compound is formed, as some high H_2S strains produced wines free from sulfide aroma. The problem of ascertaining inter-strain variability is further complicated by the fact that strains specifically selected for low hydrogen sulfide production do not necessarily remain stable to this trait (48), meaning that intra-strain variability must also be taken into account. This was corroborated in a study of three yeast strains by Sea et al. (49) who found no significant differences among yeasts; the intra-strain variability in H_2S production greatly exceeded the inter-strain variability.

Regardless, the effects of yeast strain on volatile production appear to be real. The intra-strain variability notwithstanding, Lurton et al. (50) demonstrated the efficacy of utilizing volatile compounds data in distinguishing among fifteen yeast strains indigenous to Cognac. However, the question remains, are the yeast-specific volatile differences of sufficient magnitude to influence the sensory perceptual response?

Sensory Perceptual Differences. Noble (51) outlined the difficulties in relating chemical data to sensory perceptual data. Any number of studies have endeavored to correlate chemical data to sensory response without ascertaining if the volatiles selected were indeed "aroma significant," to use Noble's phrase. Such a determination would require a human observer sniffing the effluent from a gas chromatographic run (see 26); even so, without an understanding of additive and/or synergistic effects even these data cannot define a causal relationship between chemical constituents and sensory perception.

For instance, although the total fusel alcohols average 315 mg/L in American wines (32), this concentration is not likely to have significant sensory impact, as it coincides with the detection threshold level for the most important higher alcohol, isoamyl (52). With regards to the other fusel alcohols, the threshold in wine for isobutanol alone was determined to be on the order of 500 mg/L (Rankine (52) reported a threshold value in wine of 500 mg/L and Meilgaard (53) reported a threshold value in beer of 200 mg/L). Considering that yeasts have only been observed to produce isobutanol concentrations varying by a few mg/L (31,39), and that these same yeasts produce absolute concentrations of isobutanol on the order of 20 mg/L (in agreement with the range reported by Rankine (52) of 9 to 37 mg/L), it is clear that although inter-strain differences may be statistically significant for isobutanol production, these differences are highly unlikely to be of practical sensory significance. The possible exception is isoamyl alcohol, which Rankine (52) determined to have a threshold value in wine on the order of 300 mg/L; Meilgaard (53) reported a value in beer of 70 mg/L. The yeast strains studied by Rankine (52) produced 115 to 262 mg/L of the amyl alcohols (isoamyl

+ active amyl); it is conceivable that for some persons the amount of isoamyl alcohol produced would exceed threshold. Regardless, the hedonic response related to that perception is certainly unpredictable *a priori*.

The situation with regards to diacetyl is less clear, as diacetyl production is highly variable, and as the detection threshold is cultivar dependent (*54*). Meilgaard (*53*) reported a threshold value in beer of 0.15 mg/L; detection threshold values in wine have been reported to be an order of a magnitude higher (*55*), although recent work by Martineau *et al.* (*54*) found diacetyl thresholds in Chardonnay wines averaging 0.2 mg/L, and in Cabernet Sauvignon wines averaging 2.8 mg/L. Therefore, as with the fusel alcohols, it is possible that in some instances diacetyl may be produced by yeasts in quantities exceeding the sensory threshold, although the effect on hedonic response is, once again, unpredictable.

Soles *et al.* (*42*) found the same situation to be true with regards to ester production. Meilgaard (*53*) reported a detection threshold for isoamyl acetate in beer of ~1.5 mg/L. Given that nine of the fourteen yeasts studied by Soles *et al.* (*42*) produced between 1.5 and 2 mg/L of isoamyl acetate (and all produced less than 3 mg/L), it is doubtful that the differences among the strains would be of practical sensory significance.

The difficulty is that, as Kunkee and Vilas (*39*) stated, "the modern scientific literature on this topic has not included...replicate fermentations and stringent sensory analyses." Delteil and Jarry (*31*) briefly discussed sensory aspects of the wines produced, but the details are very vague, and it's not clear whether or not these tastings were controlled in any fashion. Avedovech *et al.* (*56*) performed a careful study of five yeasts on wine aroma; unfortunately the data are confounded by *Leuconostoc oenos* strains, as the end objective of the study was to evaluate malolactic fermentation effects. Dumont *et al.* (*57*) conducted a replicated study incorporating sixteen yeast strains and three grape cultivars. However, the researchers utilized free-choice profiling techniques which only allow sensory differences to be ascertained qualitatively (*58*); there is no way to determine whether the differences observed were statistically significant. Kunkee and Vilas (*39*) ran a thorough study of five fermentations conducted in triplicate (four induced with cultured yeasts and the fifth allowed to ferment without addition) in which they found no differences in wine aroma among the strains tested.

Similar results were recently found in our laboratory, in which ten yeast strains were used to ferment Riesling juice (Edwards, C.G., Reynolds, A.G., Thorngate, J.H., unpublished data). Bench testing of the wines had initially indicated that the wines could be descriptively profiled; statistical analysis of the panel's results indicated that the sensory differences among wines were in actuality too small to be scaled–Kunkee and Vilas' (*39*) comment regarding "object lessons" is well taken. Triangular difference testing was then conducted on the wines; only two wines were found to be statistically significantly different at $p < 0.05$ (*S. cerevisiae*, V-1116 and *S. bayanus*, Champagne). However, this difference was borderline as a change in decision by only one judge would have caused the test to become insignificant using the 5% criterion. Which is to say that differences, if they did indeed exist, were subtle at best and of limited practical significance.

This is not to say that subthreshold concentrations of volatiles may not have synergistic effects; that is, exclusively requiring that compounds be present in suprathreshold concentrations may be too limiting. The combined contribution of a number of subthreshold compounds may indeed have a sensory impact. However, these effects, if they do indeed exist, were clearly not observed in the studies of Kunkee and Vilas (39) or Edwards et al. (Edwards, C.G., Reynolds, A.G., Thorngate, J.H., unpublished data).

Nor are such effects to be highly anticipated. Reed and Nagodawithana (59) cautioned that "the effect of the very large number of volatile by-products on the aroma is quite difficult to determine and evaluate." Martini and Martini (3) stated it more bluntly, concluding that "the contribution of organoleptic specificity by the yeast that causes the fermentation of a must should be considered highly improbable." And Kunkee and Bisson (23) noted that "in spite of some diverse contemporary claims there is no substantial evidence linking yeast strain with special fermentation flavours." Differences in production of volatiles certainly exist among yeast strains, but these differences have not been unambiguously demonstrated to be of reproducible, practical sensory significance.

Extrinsic Factors

Non-microbiological. If the strain-specific differences in aroma resulting from *S. cerevisiae* fermentations are indeed of minimal perceptual importance, than what accounts for the sensory variability frequently noted among fermentation lots? Kunkee and Amerine considered this question over twenty-five years ago, concluding that "the [grape] variety, the condition of the grapes used as starting material, and the conditions of the fermentation would be far more influential on quality of wine than the particular strain of *Sacch. cerevisiae*" (6). Reed and Nagodawithana writing some twenty years later drew the same conclusions regarding volatile fermentation by-products, "the flavor and aroma of a wine depends mainly on the type and quality of the must and on processing conditions" (59).

Diacetyl production is strongly influenced by winemaking conditions (34); oxygenation of the fermenting will promote oxidation of α-acetolactate to diacetyl, whereas a higher inoculation rate will reduce the diacetyl concentration. High SO_2 concentrations, and rapid removal of yeast following fermentation also lead to higher diacetyl concentrations (34).

With regards to fusel alcohol production, Kunkee and Goswell (5) noted that while yeast strain appeared to have an effect on fusel alcohol production, other factors, notably must composition, appeared to have equally important influences. For example, Berry and Watson (29) reported that added nitrogen and carbohydrates can stimulate higher alcohol production, as can increased pH (59). Various processing parameters can also affect fusel alcohol production, including agitation, aeration, and temperature (29,59).

Processing parameters also greatly affect ester production during fermentation; Killian and Ough (60) studied the effect of fermentation temperature on ester formation and retention, finding that lower temperatures preferentially favored "fruity" esters (e.g., small acetate esters) whereas higher temperatures preferentially favored long chain ethyl esters. Excess glucose also favors ester formation (29), as does removal of carbon dioxide; aeration, however, diminishes ester formation (29).

Hydrogen sulfide production is highly correlated to must composition (29). Lack of pantothenate or zinc stimulates hydrogen sulfide production, as does an excess of iron or copper (29); as free amino nitrogen decreases, hydrogen sulfide production may increase (61).

Fermentation temperature can also indirectly influence volatile concentrations; higher fermentation temperatures lead to a more rapid production of carbon dioxide, which in turn strips the wine of volatile flavor compounds (23). As volatile production peaks early in fermentation (62), increased stripping during this time would be highly disadvantageous to the final flavor.

Microbiological. It would be erroneous to conclude, however, that the sensory variability observed is due exclusively to non-microbiological sources, as the indigenous grape microflora also play an important role in fermentation (21). Fleet et al. (15) demonstrated the importance of such genera as *Hanseniaspora* and *Candida* in the early stages of natural (spontaneous) fermentation; similar findings have been reported for inoculated fermentations (63,64).

As Kunkee and Bisson (23) dryly stated, the topic of natural fermentations has been the focus of much debate, a debate which has witnessed many conflicting results (59). Amerine et al. (65) proposed that flavor-unique wines could be produced by mixed cultures under carefully controlled conditions; however, the success of such procedures has been questioned (66). It has been demonstrated that indigenous yeast cultures produce different volatile profiles (67,68), but these differences have not been rigorously demonstrated to carry over into the sensory perceptual space. Edinger and Henick-Kling (69) cautioned that any advantages to natural fermentations may be well offset by stuck fermentations and formation of off-flavors. Certainly the synergestic interactions among different yeast strains, and their effect on wine sensory properties, remain to be fully studied (23).

Biotechnological Solutions

If the current *S. cerevisiae* strains do not exhibit pronounced strain-specific effects on wine flavor, is it possible to genetically improve them to produce desired aroma components? Such improvement could be accomplished by utilizing classical hybridization, protoplast fusion, mutation/selection or genetic engineering techniques (70), presupposing that the characteristic(s) to be selected for is under the control of a single gene (59).

Thornton and coworkers (48) have used classical hybridization techniques, mating homothallic spores to heterothallic spores or cells, to improve winemaking properties, including fermentation efficiency, glycerol production and flocculation.

Similar techniques were used by Shinohara *et al.* (*71*) to develop hybrids with increased production of fusel alcohols and esters. Protoplast fusion techniques have been used to confer amylolytic activity to brewery yeasts (*22*) and ethanol tolerance to wine yeasts (*70*); Farris *et al.* (*72*) used protoplast fusion to produce hybrids with "killer factor;" that is, the ability to secrete proteinic toxins. Kunkee and coworkers (*23*) utilized a leucine auxotrophic mutant strain of *S. cerevisiae* (UCD Montrachet 522) to produce base wine for brandy production; the mutant strain produces less isoamyl alcohol, reducing the quantity of fusel alcohols in the subsequent brandy. And Thornton (*48*) discussed the progress in utilizing plasmid vectors to introduce new genes into wine yeasts; he cautioned, however, that until the yeast genome is better understood that direct gene manipulation techniques will be of limited value.

Even should genetic improvements increase flavor production, as Shinohara and coworkers (*71*) described, it must be wondered whether this increase will impart long-term improvements to the wine. Subden (*9*) contends that yeast effects are more pronounced in young wines, a point of view shared by Kunkee and Vilas (*39*) whom noted that yeast-specific effects are only unambiguously observed immediately at the end of fermentation. If so, the efficacy of such yeast improvements remains in doubt. Furthermore, even if successfully modified, genetically improved strains may grow more slowly than their counterparts, or even revert back to their parent strain (*70*). Kunkee and Bisson (*23*) also point out that any "improvements" must not affect the yeast physiology such that the wine quality is actually diminished; additionally, the improved yeasts must be physically commercially competitive with the active dry starters currently on market. Finally, it could well prove that it is not the direct formation of volatile compounds by yeasts, but rather the formation of volatile-binding macromolecules (*73,74*) or the activity of yeast enzymes (*75,76*), which merits the real attention of the yeast bioengineers.

Conclusions

Regardless of the rather pervasive opinions regarding yeast strain effects on wine flavor, the data accumulated over the past half century clearly demonstrate that these opinions are, in actuality, misconceptions regarding yeast's role in wine fermentations, at least with regards to *Saccharomyces cerevisiae*. The myths may be debunked as follows:

1) *S. cerevisiae* strains have not evolved under selective pressures in either the vineyard or the winery to optimize wine flavor. Nor do the strains, at least in the vineyard, necessarily exhibit year to year constancy.

2) *S. cerevisiae* does not provide a reliable tool for optimizing wine sensory properties. Strains may be selected for optimizing fermentation efficiency, or ethanol or sulfur dioxide tolerance, but the strain-specific production of volatiles appears to be highly variable. Much of the research has focused on the concentration of volatiles produced (which may show strain-specific

effects) without considering whether or not these differences translate themselves into the sensory perceptual space.

It is possible in any given study to demonstrate differences among yeast strains; however, these results are likely confounded with the must composition and subsequent fermentation conditions specific to that study. Since the extrinsic factors can have a far greater effect on volatile profiles than does the inoculum it is all too easy to reach erroneous conclusions regarding strain effects.

Given this situation, it would seem imperative to conduct a comprehensive study of intra-strain variability, so that winemakers could be made aware of the full range of flavor effects possible for a specific yeast strain. If this variability could be coupled to a wide range of fermentation conditions (that is, demonstrate, if possible, the dependence upon the extrinsic factors) then the winemakers would indeed have a powerful tool at their disposal. Using this knowledge the winemaker could chose the optimal strain for the conditions specific to that fermentation. Synergestic effects of multistrain (*i.e.*, natural) fermentations could then be rationally studied, as the effects of single strains would be fully characterized.

Of course, whether or not any of this information would actually yield practical benefits is unknown. It is quite possible that flavor manipulation by yeasts would necessitate such extensive compositional data of the must, and such exacting control of the fermentation conditions, as to prove unfeasible. Regardless, knowledge of the intra-strain variability might at the least demonstrate the inadequacies of depending upon yeast strains to produce specific flavors, and quell the myth once and for all.

Acknowledgments

The author would like to thank Dr. Ralph Kunkee for his generous assistance. The author is also endebted to the careful and constructive criticisms of the reviewers.

Literature Cited

1. Campbell, J. *The Power of Myth*; Doubleday: New York, NY, 1988; 237 pp.
2. Fuller, R.C. *J. Amer. Culture* **1993**, *16*, 39-45.
3. Martini, A.; Martini, A.V. In *Yeast Technology*; Spencer, J.F.T.; Spencer, D.M., Eds.; Springer-Verlag: Berlin, 1990; pp. 105-123.
4. Vaughan-Martini, A.; Martini, A. *J. Indust. Micro.* **1995**, *14*, 514-522.
5. Kunkee, R.E.; Goswell, R.W. In *Economic Microbiology*; Rose, A.H., Ed.; Academic Press: London, 1977, Vol. 1; pp. 315-386.
6. Kunkee, R.E.; Amerine, M.A. In *The Yeasts*; Rose, A.H.; Harrison, J.S., Eds.; Academic Press: London, 1970, Vol. 3; pp. 5-71.
7. Rosini, G.; Federici, F.; Martini, A. *Microbial Ecol.* **1982**, *8*, 83-89.

78

8. Martini, A.; Ciani, M.; Scorzetti, G. *Am. J. Enol. Vitic.* **1996**, *47*, 435-440.
9. Subden, R.E. In *Yeast Strain Selection*; Panchal, C.J., Ed.; Marcel Dekker: New York, 1990; pp. 113-137.
10. Phaff, H.J.; Miller, M.W.; Mrak, E.M. *The Life of Yeasts*; Harvard University Press: Cambridge, MA, 1978; 341 pp.
11. Rosini, G. *J. Gen. Appl. Microbiol.* **1984**, *30*, 249-256.
12. Constantí, M.; Poblet, M.; Arola, L.; Mas, A.; Guillamón, J. *Am. J. Enol. Vitic.* **1997**, *48*, 339-344.
13. Frezier, V.; Dubourdieu, D. *Am. J. Enol. Vitic.* **1992**, *43*, 375-380.
14. Schütz, M.; Gafner, J. *Lett. Appl. Microbiol.* **1994**, *19*, 253-257.
15. Fleet, G.H.; Lafon-Lafourcade, S.; Ribéreau-Gayon, P. *Appl. Environ. Microbiol.* **1984**, *48*, 1034-1038.
16. Querol, A.; Barrio, E.; Ramón, D. *Int. J. Food Microbiol.* **1994**, *21*, 315-323.
17. Longo, E.; Cansado, J.; Agrelo, D.; Villa, T.G. *Am. J. Enol. Vitic.* **1991**, *42*, 141-144.
18. Regueiro, L.A.; Costas, C.L.; Rubio, J.E.L. *Am. J. Enol. Vitic.* **1993**, *44*, 405-408.
19. Mortimer, R.K. *Am. Vineyard* **1994**, *3*, 12-13,19.
20. Vezinhet, F.; Hallet, J.-N.; Valade, M.; Poulard, A. *Am. J. Enol. Vitic.* **1992**, *43*, 83-86.
21. Fleet, G.H.; Heard, G.M. In *Wine Microbiology and Biotechnology*; Fleet, G.H., Ed.; Harwood Academic Publishers: Chur, Switzerland, 1993; pp. 27-54.
22. Oliver, S.G. In *Saccharomyces*; Tuite, M.F.; Oliver, S.G., Eds.; Plenum Press: New York, NY, 1991; pp. 213-248.
23. Kunkee, R.E.; Bisson, L.F. In *The Yeasts*; Rose, A.H.; Harrison, J.S., Eds.; Academic Press: London, 1993, Vol. 5; pp. 69-127.
24. Mortimer, R.K.; Romano, P.; Suzzi, G.; Polsinelli, M. *Yeast* **1994**, *10*, 1543-1552.
25. Martens, M.; van der Burg, E. In *Progress in Flavour Research 1984*; Adda, J., Ed.; Elsevier Science: Amsterdam, 1985; pp. 131-148.
26. Rapp, A. In *Modern Methods of Plant Analysis*; Linskens, H.F.; Jackson, J.F., Eds.; Springer-Verlag: Berlin, 1988, New Series Vol. 6; pp. 29-66.
27. Rapp, A.; Mandery, H. *Experientia* **1986**, *42*, 873-884.
28. Nykänen, L. *Am. J. Enol. Vitic.* **1986**, *37*, 84-96.
29. Berry, D.R.; Watson, D.C. In *Yeast Biotechnology*; Berry, D.R.; Russell, I.; Stewart, G.G., Eds.; Allen & Unwin: London, 1987; pp. 345-368.
30. Cordonnier, R.; Bayonove, C. *Conn. Vigne Vin* **1981**, *15*, 269-286.
31. Delteil, D.; Jarry, J.M. *Aust. N.Z. Wine Ind. J.* **1992**, *7*, 29-33.
32. Ough, C.S.; Amerine, M.A. *Methods for Analysis of Musts and Wines*; John Wiley & Sons: New York, NY, 1988, 2nd ed.; 377 pp.
33. Soufleros, E.; Bertrand, A. *Conn. Vigne Vin* **1979**, *13*, 181-198.
34. Martineau, B.; Henick-Kling, T.; Acree, T. *Am. J. Enol. Vitic.* **1995**, *46*, 385-388.
35. Giudici, P.; Zambonelli, C. *Am. J. Enol. Vitic.* **1992**, *43*, 370-374.

36. Romano, P.; Suzzi, G.; Comi, G.; Zironi, R. *J. Appl. Bacteriol.* **1992**, *73*, 126-130.
37. Houtman, A.C.; du Plessis, C.S. *S. Afr. J. Enol. Vitic.* **1986**, *7*, 14-20.
38. Mateo, J.J.; Jimenez, M.; Huerta, T.; Pastor, A. *Int. J. Food Microbiol.* **1991**, *14*, 153-160.
39. Kunkee, R.E.; Vilas, M.R. *Vitic. Enol. Sci.* **1994**, *49*, 46-50.
40. Ough, C.S.; In *Proceedings, University of California, Davis Grape and Wine Centennial Symposium*; Webb, A.D., Ed.; University of California Press: Davis, CA, 1982, pp. 336-341.
41. Daudt, C.E.; Ough, C.S. *Am. J. Enol. Vitic.* **1973**, *24*, 130-135.
42. Soles, R.M.; Ough, C.S.; Kunkee, R.E. *Am. J. Enol. Vitic.* **1982**, *33*, 94-98.
43. Zeeman, W.; Snyman, J.P.; van Wyk, C.J. In *Proceedings, University of California, Davis Grape and Wine Centennial Symposium*; Webb, A.D., Ed.; University of California Press: Davis, CA, 1982, pp. 79-90.
44. Goniak, O.J.; Noble, A.C. *Am. J. Enol. Vitic.* **1987**, *38*, 223-227.
45. Rankine, B.C. *J. Sci. Food Agric.* **1963**, *14*, 79-91.
46. Rankine, B.C. *J. Sci. Food Agric.* **1964**, *15*, 872-877.
47. Eschenbruch, R.; Bonish, P.; Fischer, B.M. *Vitis* **1978**, *17*, 67-74.
48. Thornton, R.J. *Crit. Rev. Biotech.* **1991**, *11*, 327-345.
49. Sea, K.; Butzke, C.; Boulton, R. *Am. J. Enol. Vitic.* **1997**, *48*, 377-378.
50. Lurton, L.; Snakkers, G.; Roulland, C.; Galy, B.; Versavaud, A. *J. Sci. Food Agric.* **1995**, *67*, 485-491.
51. Noble, A.C. In *Modern Methods of Plant Analysis*; Linskens, H.F.; Jackson, J.F., Eds.; Springer-Verlag: Berlin, 1988, New Series Vol. 6; pp. 9-28.
52. Rankine, B.C. *J. Sci. Food Agric.* **1967**, *18*, 583-589.
53. Meilgaard, M. *MBAA Tech. Quart.* **1975**, *12*, 151-168.
54. Martineau, B.; Acree, T.; Henick-Kling, T. *Food Res. Intl.* **1995**, *28*, 139-143.
55. Etievant, P.X. In *Volatile Compounds in Food and Beverages*; Maarse, H., Ed.; Marcel Dekker: New York, 1991; pp. 483-546.
56. Avedovech, R.M.; McDaniel, M.R.; Watson, B.T.; Sandine, W.E. *Am. J. Enol. Vitic.* **1992**, *43*, 253-260.
57. Dumont, A.; McDaniel, M.R.; Watson, B.T. In *Proceedings of the Twenty-Third Annual New York State Wine Industry Workshop*; Henick-Kling, T., Ed.; Cornell University, New York State Agricultural Experiment Station: Geneva, NY, 1994; pp. 42-54.
58. Lawless, H.T.; Heymann, H. *Sensory Evaluation of Food: Principles and Practices*; Chapman & Hall: New York, 1998; 819 pp.
59. Reed, G.; Nagodawithana, T.W. *Yeast Technology*; AVI: New York, NY, 1991; 454 pp.
60. Killian, E.; Ough, C.S. *Am. J. Enol. Vitic.*, **1979**, *30*, 301-305.
61. Henschke, P.A.; Jiranek, V. In *Wine Microbiology and Biotechnology*; Fleet, G.H., Ed.; Harwood Academic Publishers: Chur, Switzerland, 1993; pp. 77-164.
62. Stashenko, H.; Macku, C.; Shibamato, T. *J. Agric. Food Chem.* **1992**, *40*, 2257-2259.

63. Heard, G.M.; Fleet, G.H. *Appl. Environ. Microbiol.* **1985**, *50*, 727-728.
64. Schütz, M.; Gafner, J. *J. Appl. Microbiol.* **1993**, *75*, 551-558.
65. Amerine, M.A.; Berg, H.W.; Kunkee, R.E.; Ough, C.S.; Singleton, V.L.; Webb, A.D. *The Technology of Wine Making*; AVI: Westport, CT, 1980, 4th ed.; 794 pp.
66. Bisson, L.F.; Kunkee, R.E. In *Mixed Cultures in Biotechnology*; Zeikus, J.G.; Johnson, E.A., Eds.; McGraw-Hill: New York, NY, 1991; pp. 37-68.
67. Moreno, J.J.; Millán, C.; Ortega, J.M.; Medina, M. *J. Indust. Microbiol.* **1991**, *7*, 181-190.
68. Lema, C.; Carcia-Jares, C.; Orriols, I.; Angulo, L. *Am. J. Enol. Vitic.* **1996**, *47*, 206-216.
69. Edinger, W.D.; Henick-Kling, T. In *Proceedings of the Twenty-Fourth Annual New York State Wine Industry Workshop*; Henick-Kling, T., Ed.; Cornell University, New York State Agricultural Experiment Station: Geneva, NY, 1995; pp. 99-104.
70. Reed, G; Nagodawithana, T.W. *Am. J. Enol. Vitic.* **1988**, *39*, 83-90.
71. Shinohara, T.; Saito, K.; Yanagida, F.; Goto, S. *J. Ferm. Bioeng.* **1994**, *77*, 428-431.
72. Farris, G.A.; Fatichenti, F.; Bivulco, L.; Berardi, E.; Deiana, P.; Satta, T. *Biotech. Lett.* **1992**, *14*, 219-222.
73. Voilley, A.; Beghin, V.; Charpentier, C.; Peyron, D. *Lebensm. Wiss. Technol.* **1991**, *24*, 469-472.
74. Lubbers, S.; Charpentier, C.; Feuillat, M.; Voilley, A. *Lebensm. Wiss. Technol.* **1994**, *27*, 108-114.
75. Delcroix, A.; Günata, Z.; Sapis, J.-C.; Salmon, J.-M.; Bayonove, C. *Am. J. Enol. Vitic.* **1994**, *45*, 291-296.
76. Zoecklein, B.W.; Marcy, J.E.; Jasinski, Y. *Am. J. Enol. Vitic.* **1997**, *48*, 397-402.

Chapter 7

Seasonal Variation in the Production of Hydrogen Sulfide During Wine Fermentations

Kevin Sea, Christian Butzke, and Roger Boulton

Department of Viticulture and Enology, University of California, Davis, CA 95616–8749

The production of hydrogen sulfide during wine fermentation was measured during two seasons (1995 and 1996) using a direct headspace sampling procedure and gas chromatography. Total sulfide formation was correlated with individual amino acid, free amino nitrogen, total amino nitrogen and various ratios of these variables in the starting juices using the method of principal component analysis. The major factors in juice composition that were correlated with total sulfide formation differed between the seasons. Hydrogen sulfide formation in 1995 was most strongly correlated with higher levels of glycine, β-alanine and citrulline while in 1996 it was with phenylalanine, leucine, alanine, free amino nitrogen and methionine. When the data for both seasons was combined, the sulfide formation was again most strongly correlated with glycine, β-alanine and citrulline.

The formation of hydrogen sulfide (and other organic sulfides) by yeast during wine fermentations continues to be one of the major defects in modern winemaking. There can be a number of associated sulfur-containing products in young wines such as methyl and ethyl thiols, their corresponding thioacetates, dimethyl sulfide and sulfite. Hydrogen sulfide is often the most obvious component since it has a low sensory threshold concentration above wine (50 to 80 ug/L) *(1)* and its characteristic "rotten egg" smell is generally recognized even when the thiols are present. The evolution of large quantities of carbon dioxide during fermentation and the relatively fast diffusivity of hydrogen sulfide often lead to it being quickly distributed throughout a fermentation cellar once its formation has begun. The thiols pose a particular danger in that they are easily oxidized to their corresponding disulfides which have much higher sensory threshold concentrations, only to be cleaved by sulfite in the absence of oxygen, usually several months later and often after bottling, reforming the thiols at supra-threshold concentrations *(2)*. The thioacetates pose an even more challenging situation since they are far less volatile, often present at subthreshold levels yet they can undergo hydrolysis at wine pH to yield thiols at supra-threshold concentrations months after the ethanol fermentation has been completed.

There is presently no known prevention strategy for the formation of these compounds during wine fermentations. Of the hundreds of yeast strains isolated from many wine-producing areas of the world, no consistently low sulfide producing strains are commercially available. While there have been dramatic developments in the understanding of yeast genetics due to the recently completed mapping of the Saccharomyces cerevisiae genome, this has yet to provide sufficient information to enable the development of strains of wine yeast that produce significantly less or essentially no hydrogen sulfide during fermentation. Our studies over several seasons (1990 to 1996) have consistently shown that while yeast strain is a factor in the formation of sulfides in any particular juice or medium, when evaluated over several juices, the results average out with no strain being consistently better than others in terms of lower sulfide production.

The largest variation in sulfide production continues to be associated with the variations in juice or medium composition. While there have been several suggestions of the factors preventing this production, such as the enhancement of methionine or ammonium or the free amino nitrogen content, the deferment of adding sulfite until after fermentation and the introduction of oxygen when sulfide appears, none of these have been found to be generally successful for the conditions that we have observed over many seasons with juices from California. Many of the effects demonstrated in model medium with single sulfur sources cannot be replicated in actual juices and a number of questions remain regarding compositional factors in juices that contribute to the sulfide formation. There is considerable disagreement in the literature regarding the causal factors even in juice studies and the absence of any universal nutrient addition or supplement that prevents sulfide formation is strong evidence that the situation is more complicated than is generally thought, perhaps involving multiple interactions .

The interpretation of peaks and rates of formation need to be carefully examined since the rates of change in concentration vary due to both the rate of sulfide formation and the rate of carbon dioxide evolution. Increases in concentration can (and do) occur under conditions of low gas evolution, even when the rate of sulfide formation is constant. This becomes especially important when a decrease in the concentration in mid fermentation can be due to the gas evolution rate alone. In some literature (and perhaps in practice) this leads to the mistaken interpretation that the formation has diminished. The same logic can be applied to attribute high concentration at the beginning and end of fermentation to higher formation rates when these are the points of the lowest gas evolution rates.

While the hydrogen sulfide formed during fermentation can be easily removed with the addition of copper sulfate, the international legal limit of 0.2 mg/L residue in wines and the natural complexing and binding of copper ions in wines often prevent suitable additions from being made in practice. Of more concern is the secondary thiols and thioacetates that are produced with hydrogen sulfide (3) and the inability of copper to remove them or their oxidized disulfide forms.

The natural formation and excretion of low levels of sulfite, sulfides and thiols is a general feature of the ethanol fermentation by Saccharomyces cerevisiae under winemaking conditions. Typical levels of yeast-derived sulfite in wines range from 10 to 30 mg/L (*4*). This is usually present in a bound form as acetaldehyde sulfonate and not of a concern to the sensory qualities of wine. The formation of trace headspace levels of hydrogen sulfide (1 to 10 ug/L) during fermentation and that remaining after the stripping action of the carbon dioxide released by the fermentation are readily detected and generally considered offensive and a winemaking defect.

The increased formation hydrogen sulfide during fermentation can be the result of elemental sulfur residues on grapes that have been sprayed with sulfur suspensions to prevent mildew and mold growth during maturation (*5-8*). The level of such residues need to be above 4 mg/L to cause problems with hydrogen sulfide formation, even though there are a number of studies with grape juice that have used levels well above those found in practice (*5,6*).

The among the first compositional effects on sulfide formation was the suggestion that low levels of free amino nitrogen (FAN) caused increased formation of hydrogen sulfide (*9*). The suggestion that low amino nitrogen was a cause stemmed from the thinking that under such conditions, yeast might produce an extracellar protease that would hydrolyze peptides and proteins, causing hydrogen sulfide formation when disulfide bonds were cleaved. Closer examination of this study shows that while sulfide formation was often lowered by such additions, it was rarely eliminated. Further, at any given FAN level, there continued to be significant variation in sulfide formation and that the variation in juice FAN could only account for a lesser part of the total formation. Other studies (*10,11*) have proposed that the depletion of amino nitrogen was a condition that could be related to the onset of hydrogen sulfide formation due to cellular metabolic changes, not the extracellular protease picture. Both of these results seem to be quite yeast strain specific and not generally applicable to Californian conditions. The free amino nitrogen is exhausted in almost every wine fermentation and yet sulfide formation varies from juice to juice. There is also the possibility that the effect of nitrogen additions merely extend the growth period. This shift in timing can result in the sulfide formation coinciding with the peak of carbon dioxide release wherein its concentration is diluted and the problem is less obvious. In addition, what was an early appearance caused by a nutrient deficiency during rapid yeast growth can be moved into a later one and this has generally been taken in some studies to mean that it has eliminated the sulfide formation.

Studies in brewing have shown a requirement in some yeast strains for trace levels (50 to 75 ug/L) of pantothenate in order to suppress sulfide formation (*12*). While one study (*13*) found that the Montrachet strain has such a requirement and others (*14*) have shown it to be a common requirement in wine yeast cultures, there is little available data on other commercial strains presently used in the California. In some studies the addition of pantothenate to defined media in a standard practice, but it is not generally found in commercial yeast foods and the existence of such deficiencies in some commercial fermentations cannot be ruled out as a source of sulfide formation.

The role of methionine and cysteine in juices has received some attention as possible controlling factors in the sulfate reduction pathway. There have been some suggestions that deficiencies of these amino acids could be a cause sulfide formation and completely contrary view that higher than usual levels were a contributing factor. The addition of high levels (1 g/L) of methionine to grape juice (*15,16*) resulted in the formation of an unpleasant smelling and dangerous component 3-(methylthio)-propanol (*16*) and hydrogen sulfide production (*15*).

Surprisingly, there have been only a few studies of the broader role of grape juice composition on sulfide formation and the view that it is the mixture composition that is the controlling influence is not widely accepted as yet. From an uptake and transport point of view the ability of the cell to take in certain components will be more related to the external competition with components for the sites of transporting enzymes than with those in certain biochemical pathways within the cell. The other reason for trying to identify the compositional characteristics of juices that lead to sulfide formation is the be able to recognize juice that are likely to have sulfide formation before the fermentation begins. It would also be useful in any longer term attempts to relate this to viticultural conditions. The correlation of individual amino acids and groups of them as well as the consideration of particular concentration ratios has been a central theme in our studies during the past six years. The poor correlations that we have observed with FAN points to the rather narrow range of conditions under which most studies have been conducted or analyzed. The work described below is a statistical analysis of seasonal comparison of more than 40 juices from the 1995 and 1996 seasons.

Experimental Methods

Source of Grapes: The grapes used for this study were taken from coastal vineyards of California, often based on vineyards which had exhibited sulfide problems in the prior year. The experimental setup was limited to a maximum of six sample per week with the sample number varying due to availability and harvest dates of the vineyards concerned. The grapes were gathered in 50 cluster samples, taken from evenly spaced rows and from vines on alternating sides, with a rotation of position (close to trunk, middle or end of cordon, alternating from the right side or the left side cordon). In 1995 there were 31 samples (14 white, 17 red) while in 1996 there were 12 samples (1 white, 11 red).

Fermentation Conditions: All juices had additions prior to fermentation in order to eliminate deficiencies from being a factor in hydrogen sulfide formation. These included 120 mg N/L in the form of diammonium phosphate (DAP), 50 mg/L SO_2, 75 ug/L pantothenate, 2 ug/L biotin and 75 ug/L thiamin. All fermentations were inoculated with 240 mg/L of active dry wine yeast (Fermivin), that had been reactivated in 35°C water. All fermentations were conducted in duplicate at 25°C, in temperature controlled, constantly stirred (100 rpm), fermentors (Applikon) using 500 mL of white juice or 300 mL juice of red juice plus the corresponding amount of skins and seeds.

Analysis Methods: Hydrogen sulfide was determined using gas chromatography and a direct headspace sample injection onto a combination capillary column (*17*). Sulfur concentrations were determined by reference to external standards of hydrogen sulfide using flame photometric detection . The total amount of hydrogen sulfide formation was estimated by integration of the daily concentrations, weighted by the daily carbon dioxide evolution. The composition of individual amino acids in the starting juices was analyzed and computations of the free amino nitrogen and assimilable nitrogen and a number of other measures were made.

Statistical Analysis: The juice composition, as individual amino acids, free amino nitrogen and ratios of certain amino acids and groups of them were entered into a principal component analysis (PCA) using the SAS (Cary, NC) statistical software. The ratios considered were based on the thinking that relative proportions within the amino pool, rather than absolute quantities between alternative substrates might be related to the level of sulfide formation.

Discussion

The formation of hydrogen sulfide during juice fermentation has been shown to have two major stages (6, 7) and these are sometimes referred to as stage I and stage II respectively.. The first stage is generally associated with nutrient deficiencies incurred during the growth phase of the yeast which is generally completed by mid-fermentation. The second stage of formation usually occurs at the end of fermentation when low sugar levels are reached and this is associated with non-growing yeast activity. The hydrogen sulfide formed by the presence of high levels of elemental sulfur (> 4 mg/L) generally appears primarily during stage II, (6, 7) that is at towards the end of active fermentation and carbon dioxide release. In the present studies, the emphasis is on total sulfide formation and this is generally that of second stage formation since it is our experience that this is the most common occurrence in commercial winemaking today. It is this formation which is quite variable from juice to juice and is not lowered by the use of ammonium salts. The addition of vitamins and ammonium salts prior to fermentation make our studies of the stage II kind and thereby make our total sulfide formations to be measures stage II formation quantities. The formation of hydrogen sulfide even with such additions demonstrates why such additions cannot be used to address this problem in commercial practice.

The statistical analysis of juice composition included the individual amino acid levels as well as various ratios such as those between threonine, serine, aspartic acid and methionine with FAN and with each other. The juice samples from the 1996 harvest are considerably richer in both total and amino nitrogen levels, almost double those of the 1995 harvest (Table 1). The 1996 samples are also fewer in number and almost exclusively red grapes. A typical level for free amino nitrogen content for normal fermentation of grape juices is in the range 120 to 150 mg N/L and on this basis, the 1995 samples can be considered "high" and those in 1996, "very high" in terms of amino nitrogen. The formation of hydrogen sulfide under these conditions cannot be attributed to low levels of free amino nitrogen. Similarly, deficiencies in either pantothenic acid or thiamin cannot be causing this production due to their addition at above requirement levels, prior to fermentation.

The major factors in juice composition that are correlated with sulfide

Table 1. General Seasonal Summaries[1]

1995 (n=31)

Juice Total Nitrogen Content	240, (138) mg/L
Juice FAN	184, (100) mg/L
H2S Produced	74, (128) ug/L

1996 (n=12)

Juice Total Nitrogen Content	404, (95) mg/L
Juice FAN	349, (53) mg/L
H2S Produced	50, (56) ug/L

[1] The values shown are the means with the standard deviations in parentheses.

Table 2. General Correlations With Hydrogen Sulfide Formation[1]

1995 Harvest (n=31)

Glycine +0.43
β-Alanine +0.43
Citrulline +0.37

Free Amino Nitrogen -0.23
Total Nitrogen -0.13
Methionine +0.10

1996 Harvest (n=12)

Phenylalanine +0.71
Proline +0.62
Leucine +0.58
Alanine +0.55
Free Amino Nitrogen +0.54
Methionine +0.49

1995 and 1996 Harvests Combined (n=43)

Glycine +0.43
β-Alanine -0.40
Citrulline +0.36
Methionine/FAN +0.36

Free Amino Nitrogen -0.20
Proline -0.32

[1] The standard linear correlation coefficient

Table 3. Principal Component Analysis - Loading of Factors 1 and 2

1995 Harvest

Factor 1 (41%)
Total Nitrogen +0.97
Valine +0.94
Free Amino Nitrogen +0.93
Phenylalanine +0.93

Factor 2 (14%)
Glycine +0.59
Hydrogen Sulfide +0.47

1996 Harvest

Factor 1 (46%)
Free Amino Nitrogen +0.95
Total Nitrogen +0.90
Methionine +0.90
Methionine/FAN +0.79

Factor 2 (19%)
Methionine/Arginine +0.79
Tyrosine +0.74
Arginine -0.74

1995 and 1996 Harvests

Factor 1 (36%)
Total Nitrogen +0.96
Valine +0.95
Free Amino Nitrogen +0.94
Leucine +0.93

Factor 2 (14%)
Methionine/Arginine +0.70
Arginine +0.67
Alanine +0.64

formation are glycine, β-alanine and citrulline (in 1995) and phenylalanine, proline, leucine, free amino nitrogen and methionine (in 1996). When the data of the two seasons are combined, the factors become glycine, β-alanine, citrulline and the ratio of methionine to free amino nitrogen. The effect of the free amino nitrogen is not significant in 1995, significant and positively correlated in 1996 and insignificant when the data of the two seasons are combined. It should be pointed out that the correlation coefficient between hydrogen sulfide formation and juice FAN in Vos and Gray's much referenced study (9) was -0.59, with n=104 juices based on simple correlation but was considerably weaker -0.34 when a partial correlation was performed including other measures. Similarly, the lack of any strong relationship between sulfide formation and initial juice ammonia content in the present studies, indicates there is little basis for using this measure in preventative strategies such as the addition of diammonium phosphate.

The correlation of proline with hydrogen sulfide is probably due to its strong linkage with total nitrogen in general since the 1996 harvest studied mostly red grapes and Cabernet Sauvignon has been shown to have relatively high levels of proline. The proline is not used extensively by yeast during wine fermentations and therefore it has not been included into the free amino nitrogen values in these studies. The correlation with methionine is interesting, especially for the season in which the nitrogen levels are very high and most of the grapes were red. It is expected that juices that are higher in assimilable nitrogen would support yeast growth further into the fermentation and in the process, produce higher concentrations of cells. The cell mass was not measured in these fermentations due to the difficulty in doing so in the presence of grape pulp and skins. The high proportion of red grapes used in these fermentations is in response to suggestions from participating wineries of their more problematic vineyards. It has yet to be established whether there is a strong relationship between cell mass produced during fermentation and the corresponding stage 2 (or in fact, total) sulfide formation. A relationship between higher final cell mass and higher initial nitrogen content does exist, even in juices that have been produced by ammonium salt additions (18) as well as those with naturally occurring amino nitrogen (19).

The principal component analysis shows that scatter between the juices (open symbols white fermentations, solid symbols for red fermentations) of the two seasons (Figure 1 and Figure 2) as well as the major correlating variables previously noted. The diagram for the combined data (Figure 3) shows the stronger importance of the phenylalanine and proline of the red grapes. There is no obvious pattern with either white or red fermentations while those producing more sulfides are clearly correlated with the variables mentioned in the general correlations.

There are two limitation of the PCA approach as presently used to look for relationships between sulfide formation and the composition of the corresponding juices. The first is that there is no accommodation of any saturation effect at higher concentrations of any of the measures. The saturation phenomenon is quite common in transport systems in most organisms and above the saturating concentration, there is no further increase in rates with increasing concentration. This effect could be accounted for by using a transforming function of the kind $[C]/[Km + [C]]$, where

90

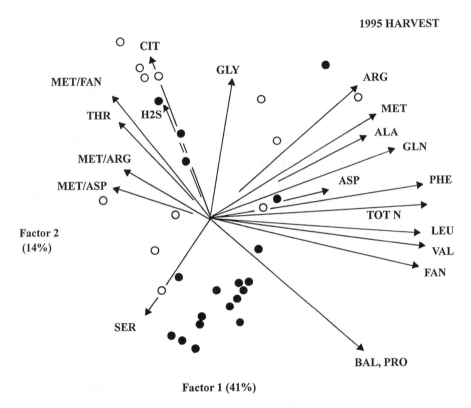

Figure 1. Principal component diagram for the 1995 harvest, Factors I and II.

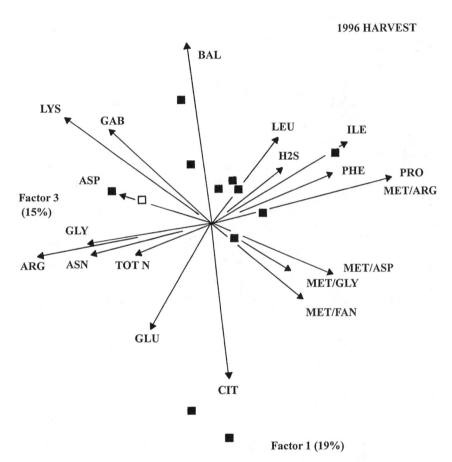

Figure 2. Principal component diagram for the 1996 harvest, Factors I and II.

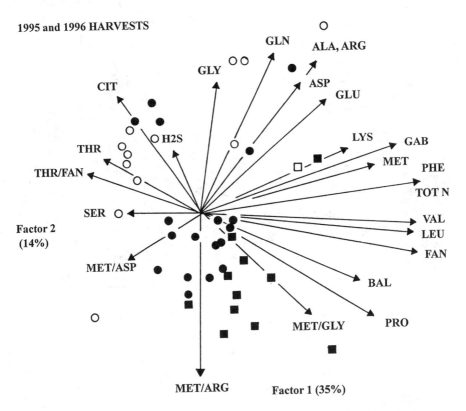

Figure 3. Principal component diagram for the combination of 1995 and 1996 harvests, Factors I and II.

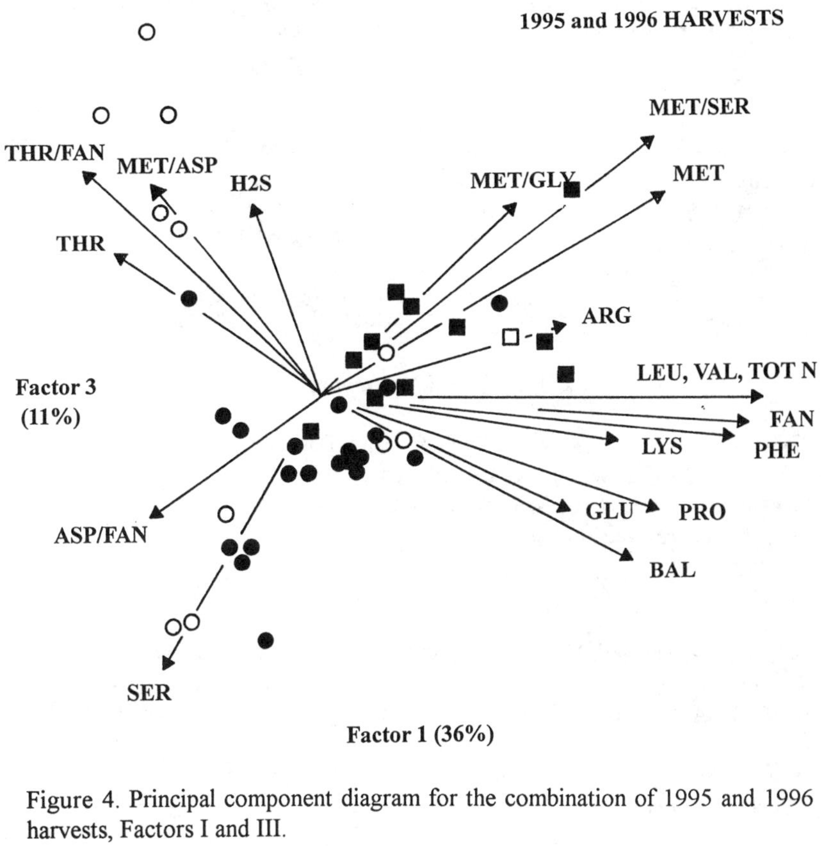

Figure 4. Principal component diagram for the combination of 1995 and 1996 harvests, Factors I and III.

Km is an apparent affinity constant, rather than the concentration [C] alone, used in this study. Work is in progress on the saturation approach to these correlations. The second limitation is that at the point of sulfide formation, generally at the end of these fermentations, there is little relationship between the medium composition that is present and that of the juice from which it was derived. These studies are the first to statistically investigate the relationship between sulfide formation during fermentation and juice composition and the use of initial composition is based on the premise is that there might be something characteristic about the juice composition that can be recognized. Such a characteristic would enable juices to be screened during vineyard sampling and perhaps modified prior to fermentation so as not to form significant hydrogen sulfide during fermentation.

Conclusions

These studies demonstrate the considerable differences observed between seasons in the formation of hydrogen sulfide during the later stages of fermentation.

Hydrogen sulfide formation was most strongly associated with juices that were higher in glycine and citrulline levels, lower in β-alanine levels and with higher ratios of methionine to FAN.

There was no significant correlation of hydrogen sulfide formation with either the free amino nitrogen or total nitrogen levels.

Acknowledgments

This work has been funded in part by grants from Baritelle Vineyards, Beringer Wine Estates, E & J Gallo Winery, Gist Brocades, Robert Mondavi Winery, Stag's Leap Wine Cellars, Sutter Home Winery, Trefethen Vineyards, William Hill Winery and the American Vineyard Foundation.

Literature Cited

1. Wenzel, K.; Dittrich, H. H.; Seyffardt, H. P. and Bohnert, J. *Wein Wissenschaft,* 1980, *35*:414-420.

*2.*Bobet, R. A.; Noble, A. C. and Boulton, R. B. *J. Agric. Fd. Chem.* 1990, *38*,449-452.

3. Ruhaut, D.; Kurbel, H. and Dittrich, H. H. *Vitic. Enol. Sci.* 1993, *48* ,214-218.

4. Weeks, C. *Am J. Enol.* 1969, *20,* 32-39.

5. Rankine, B. C. *J. Sci. Fd. Agric.* 1963, *14*, 19-91.

6. Schutz, M. and Kunkee, R. E. *Am. J. Enol. Vitic.,* 1977, *28*, 137-144.

7. Thomas, C. S.; Boulton, R. B.; Silacci, M. W. and Gubler, W. D. *Am. J. Enol. Vitic.* 1993, *44*:211-216.

8. Ruhaut, D. In *Wine Microbiology and Biotechnology*, Fleet, G.H., Ed.; Harwood Academic Publishers, Chur, Switzerland. 1993, 183-223.

9. Vos, P. J. and Gray, R. S. *Am. J. Enol. Vitic.* 1979, *30*, 187-197.

10. Jiranek, V. and Henschke, P. A. *Aust. Grapegrower & Winemaker,* 1991, 27-30.

11. Henschke, P. A. 11th International Oenology Symposium, Hungary. (1996).

12. Wainwright, T. *J. Applied Bacteriol*, 1971, *34*, 161-171.

13. Davenport, M. *The effects of vitamins and growth factors on growth and fermentation rate of three active dry wine yeast strains*. M.S. Thesis, 1985, University of California, Davis, CA.

14. Eschenbruch, R.; Bonish, P. and Fisher, B. M. *Vitis* 1978, *17*, 67-74

15. Navara, A. and Minarik, E. *Wein Wissen.* 1974, *29*, 208-215

16. Mueller, C. J.; Kepner, R. E. and Webb, A. D. *Am. J. Enol. Vitic.* 1971, *22*, 156-160.

17. Park, S. K. *Factors affecting formation of volatile sulfur compounds in wine*. Ph.D. Thesis, 1993, University of California, Davis, CA..

18. Agenbach, W. A. Proc. *S. Afric. Soc. Enol. Vitic.* 1977, 66-87.

19. Montero, F. and Bisson, L. F. *Am. J. Enol. Vitic.* 1991, 42, 199-208.

Chapter 8

What Is "Brett" (*Brettanomyces*) Flavor?: A Preliminary Investigation

J. L. Licker, T. E. Acree, and T. Henick-Kling

Department of Food Science and Technology, Cornell University, New York State Agricultural Experiment Station, Geneva, NY 14456

Barnyard, horse sweat, Band-aid, burnt plastic, wet animal, wet leather: all have been used to describe an aroma or flavor characteristics in some wines deemed "Bretty". The organisms cited for the production of this character are the yeasts of the genus *Brettanomyces* and *Dekkera*. In the literature, 4-ethyl phenol and 4-ethyl guaicol are the identified volatile phenolic compounds associated with this off-odor in wine. Included in this report is a review of "Brett" flavor and results from our recent study on wines identified by their respective wine makers as having "Brett" character. In wines with "Brett" character, sensory profiles showed an increase in plastic odors and a decrease in fruit odors. Analysis by gas chromatography-olfactometry (GCO) revealed two predominate odor-active compounds: isovaleric acid and a second unknown compound; other identified odor-active compounds included guaiacol, 4-ethyl guaiacol, 4-ethyl phenol, 2-phenyl ethanol, ß-damascenone, isoamyl alcohol, ethyl decanoate, cis-2-nonenal and trans-2-nonenal. Using the technique CharmAnalysis for GCO analysis, along with gas chromatography-mass spectrometry (GC-MS), odor-active compounds were identified by their respective Kovàts retention indices.

Literature Review

The Beginning of "Brettanomyces". N. Hjelte Claussen, then director of the Laboratory of the New Carlsberg Brewery, in Copenhagen, Denmark, introduced the word "Brettanomyces" at a special meeting of the Institute of Brewing in April 1904 (1) . Claussen proved that a type of English beer known as stock beer underwent a slow secondary fermentation after the completion of the primary fermentation. The secondary fermentation was induced by inoculating the wort with a pure strain of Brettanomyces: a non-*Saccharomyces, Torula*-like asporogeneous (non-spore forming) yeast. The flavors produced during the secondary fermentation were characteristic of the strong British beers of that time. Claussen chose the name "Brettanomyces" for the close connection between the yeast and the British brewing industry.

Table 1. Yeast names adapted into English[a]

Saccharomyces "sugar fungus"		
saccharo	sugar	Greek -- saccharon
myces	fungus	Greek -- myketes

Brettanomyces: "British brewing industry fungus"		
brettano	British brewing industry	(1)
myces	fungus	Greek -- myketes

[a] Adapted from Riesen (2)

In 1903 Claussen obtained a patent in England for his process of adding *Brettanomyces* yeast "to impart the characteristic flavour and condition of English beers to bottom-fermentation beers and for improving English beers" (3) . At that time it was unknown how the wine-like flavor developed in British beers. Brewers used the method developed by Hansen in 1883 for the inoculation of pure yeasts in bottom fermented beers; however, they were unsuccessful in their attempts to use the method to recreate the flavors of well-conditioned top fermented English stock beers. These were stored in cask, vat or bottle for more than a week after racking.

Unfortunately for Claussen's discovery, the strength of British beers began to decline, in large part due to excise tax increases (4-7) . Low attenuated beers that forgo storage after racking (running beers) replaced the stock beers along with the associated flavor characteristic of this British national beverage (7)   Claussen (1) noted a beer must reach a certain degree of attenuation to receive the benefits of a "pure flavoured product"; otherwise, the low attenuated beer "thus infected (with *Brettanomyces*) possesses a peculiar impure and sweet mawkish taste, whilst at the same time an English character becomes apparent to the nose and a very similar impure taste is the result" (1) .

***Brettanomyces* morphology and physiology.** In 1940 the first systematic study of *Brettanomyces* yeast was conducted by M. T. J. Custers (8) . In his investigation he characterized the morphology and physiology of 17 strains obtained mainly from beer. They included beer strains donated to the CBS by Claussen (1) , Kufferath and van Laer (9) , Shimwell (6) , and the Scandinavian Brewery Laboratory in Copenhagen, as well as new isolates from Belgian lambic beer, English stout & ale. The only strain not of brewery origin was from a 1930 French wine fermentation isolated by Krumbholz and Tauschanoff (10) of Geisenheim.

Custer (8) determined that all strains had several characteristic properties in common: ogive (pointed arch) cells, asporogeneous, short-lived, delayed growth on malt extract and malt agar, and production of large amounts of acetic acid under aerobic conditions. He confirmed the distinction of the genus *Brettanomyces* Kufferath et van Laer and the two species *Brettanomyces bruxellensis* and *Brettanomyces lambicus*; he distinguished two additional species: *Brettanomyces claussenii* and *Brettanomyces anomalus*.

Using *Brettanomyces claussenii*, Custers showed glucose fermentation is inhibited under anaerobic conditions. Glucose was fermented more rapidly under aerobic than anaerobic conditions. He named this inhibitory effect a "negative Pasteur effect". Aerobic conditions activated the fermentation of glucose to produce ethyl alcohol, carbon dioxide, and "a considerable amount of acetic acid". Only ethanol and carbon dioxide were produced under anaerobic conditions; acetic acid was not.

In 1961 Wikén (11) showed evidence of a negative Pasteur effect as characteristic of all yeast in the genus *Brettanomyces*. In 1966 Scheffers (12) described this inhibition of alcoholic fermentation under anaerobic conditions as a consequence of the net reduction of NAD^+ to NADH in *Brettanomyces* yeast cells: he called this a "Custers effect".

Burk (13) provided an extensive history of the literature up to 1939 on the mechanism hypotheses of a Pasteur effect in biological systems. For further confirmational and mechanistic work on a Custers effect in *Brettanomyces*, the articles by Scheffers (14), Carrascosa (15), and Wijsman (16) should be consulted.

Brettanomyces in Wine Production

Early wine research. In 1930 Krumbholz and Tauschanoff (10) isolated the yeast *Mycotorula intermedia* from a French grape must; Custer (8) reclassified it as *Brettanomyces bruxellensis*. In 1911 from Wädenswil, Osterwalder (17) isolated and identified the yeast *Monilia vini* in Swiss apple wine; Schanderl (18) and Schanderl & Draczynski (19) presumed it to be a *Brettanomyces* species; based on the physiological evidence of Osterwalder, van der Walt & van Kerken (20) characterized it as *Brettanomyces intermedius*.

In the early 1950's Schanderl (18) and Schanderl & Draczynski (19) of Geisenheim reported the first uncontested occurrence of *Brettanomyces* in bottled wine (21) ; the isolation was from a German sparkling wine. Florenzano (22) isolated four *Brettanomyces* strains in red wines from the northeast Italian city of Padua; he (23) observed additional isolates of this genus in musts and wines from the southeast Italian seaport of Bari and from Padua. The first report from France was in 1955 when Barret (24) isolated *Brettanomyces* species from 'yellow' wines of the Jura: white wines from the Savagnin grape, cask-aged for a minimum of six years (25) . Other researchers reported *Brettanomyces* isolates in wine from French wine regions: Galzy & Rioux (26) in wine from Midi; Domercq (27) and Peynaud & Domercq (28) in red and white wines from Gironde {Graves}, {St. Emilion}, {Médoc}; Chatonnet (29) in red wines from Bordeaux {Graves}, Haut-Medoc Margaux, Montagne-St. Emilion, {Pessac-Léognan}, {Madiran}; and Larue (30) in wine from St. Emilion and Médoc.

In a five-part study from 1958-1961, van der Walt & van Kerken (20, 21, 31-33) found *Brettanomyces intermedius* was mainly responsible for yeast hazes in dry white wines from South Africa. Zyl (34) also conducted an extensive study on turbidity in South African dry wines caused by *Brettanomyces*. Out of 480 turbid wines, 32 wines representing 16 wineries showed *Brettanomyces* growth.

In addition to Germany, France, and South Africa, *Brettanomyces* species were also reported in wine from other winemaking areas of the world: Italy (35) ; Brazil (36) ; Uzbekistan (37) , Spain (38-40) , Portugal (41) , New Zealand (42) , Great Britain (43) , Australia (44) , and the United States (29, 30, 45, 46) .

***Brettanomyces* within the winery.** Peynaud & Domercq (28) noted that viable *Brettanomyces* were present on the walls and in the soil of damp cellars in French wineries. Later Peynaud (47) advised, "the winemaker should imagine the whole surface of the winery and equipment as being lined with yeasts". The early South African researchers (31-34) established the importance of sanitary control at the reception of the grapes into the winery to control the growth of wild yeasts. *Brettanomyces* cultures were isolated from 13 of 53 samples from winery equipment collected in five South African wineries (33) . However none of these yeasts was isolated from winery floors, walls, or equipment cultured in any of the 15 New Zealand wineries investigated by Wright & Parle (42) during the 1971 vintage, even though "considerable numbers" were present in fermentations.

***Brettanomyces* in must.** Researchers isolated *Brettanomyces* in fermenting grape must from around the world: France (10, 27) ; Germany (18) ; Italy (23, 48) , South Africa (33) ; Uzbekistan (37) ; New Zealand (42) ; and Spain (40, 49) . *Brettanomyces* populations were rarely found to be the predominate species in the microflora of fermenting musts, although some were detected. Domercq (27) detected two *Brettanomyces* cultures out of 80 red; no cultures were isolated out of 38 white grape musts sampled from French wineries.

In preliminary studies, van der Walt & van Kerken (33) were unable to isolate any *Brettanomyces* cultures from fermenting must using methods developed by Domercq (27) . They developed a selective medium which included the addition actidione and sorbic acid. A few years earlier, Beech & Carr (50) conducted a survey of inhibitory compounds and found 50 mg/L actidione plus 500 mg/L sorbic acid inhibited the growth of all yeast except *Brettanomyces* and *Trigonopsis*. Using this medium, van der Walt & van Kerken (33.) isolated one culture out of 10 white grape musts sampled from 10 South African wineries. *Brettanomyces* were an uncommon contaminant in fermenting musts, and no cultures were isolated from vineyard grapes (33) .

In contrast to van der Walt & van Kerken's findings, *Brettanomyces* were isolated consistently from fermentations in 6 of 15 New Zealand wineries, and occasionally from 4 of the 15 (42) . In total they were isolated in 33 of 124 fermenting white and red musts in New Zealand. The New Zealand researchers found the methods of van der Walt & van Kerken (33) to be completely inadequate for *Brettanomyces* growth, as well as media containing 50 mg/L actidione, 500 mg/L potassium sorbate or 0-16% ethanol. Wright and Parle developed their own selective media for the rapid growth of *Brettanomyces* spp.: 20% sucrose, 0.7% $(NH_4)_2HPO_4$, 0.4% $(NH_4)_2SO_4$, 0.4% K_2SO_4, 0.3% yeast extract acidified to pH 4 with tartaric acid.

Brettanomyces **detection.** Kunkee and Amerine (51) wrote of the problems associated with yeast detection: "The major problem of studies on yeast ecology results from the methods used for isolating the microflora. So many different techniques have been used that comparisons of frequency of occurrence must be made with caution". One compendium entitled "Media and Methods for Growing Yeasts" provides reference to the variety of techniques (52) .

Van der Walt & van Kerken (33) were initially unable to isolate *Brettanomyces* from musts and winery equipment using customary media and techniques. It can take a week or longer before colonies are visible, a far longer incubation time compared to *Saccharomyces* and other yeast. Custer (8) was first to note the characteristically slow growth of the genera. *Saccharomyces, Kloeckera, Metschnikowia, Pichia, Candida,* or "wild" yeast develop earlier in culture thereby hindering detection of *Brettanomyces.* Early mold development inhibits detection by covering the media surface. *Brettanomyces* may go undetected; plates may be discarded before colonies develop. Addition of 100 mg/L cycloheximide (actidione), 10 mg/L thiamin, and 0.5% calcium carbonate to media aids in the selective detection of *Brettanomyces* yeast (53) .

Brettanomyces **and barrels.** Another problem of *Brettanomyces* detection in the winery is large variations in barrel to barrel populations. In a 45 week barrel sampling study of stored Cabernet Sauvignon wine, Blazer and Schleußner (54) determined it was necessary to stir barrels before plating to acquire accurate cell counts of *Brettanomyces.* Measured populations increased after stirring -- in some cases by 10-fold or more; in others, detection depended on stirring.

Fugelsang (55) stated that wood cooperage is the most frequently cited source of *Brettanomyces* within the winery. In 1990, Van de Water (The Wine Lab, Napa, California) reported that in hundreds of wineries from across the United States, *Brettanomyces* infection within a winery could be traced to purchased wooden cooperage used previously for red and infrequently for white wine production (45) . Even new barrels are suspected of having a stimulatory effect on the growth of *Brettanomyces:* some of the species can assimilate cellobiose and thrive on these fragments of cellulose in new barrels (56). Wineries are encouraged by some enologists in the United States to destroy *Brettanomyces*-infected barrels to avoid further contamination within the winery (56, 57) .

Another problem is barrel to barrel variations within the same lot. In a recent two-year Californian winery investigation (57, 58) , new oak, stainless steel, and previously *Brettanomyces*-infected barrels were filled with Cabernet Sauvignon inoculated with 4 cells/mL of a *Brettanomyces* culture. Similar barrels were filled with sterilized wine. All of the wooden barrels (new & old containing inoculated & sterile wine) developed *Brettanomyces* populations. *Brettanomyces* populations were undetectable in all of the stainless steel barrels. They suggested *Brettanomyces* growth may be due to favorable growth conditions (oxygen, nutrients, or both) than to direct infection from old barrels.

Sulfur dioxide treatment. According to Chatonnet (59) , the only way to limit the growth of *Brettanomyces* in red wines aged in oak barrels is to maintain a sufficient concentration of free sulfur dioxide (SO_2) throughout the aging process. At least 7 g/barrel of SO_2 gas should be used to disinfect empty barrels. Filled wine barrels should receive 20 to 25 mg/L of free SO_2, 30 to 35 mg/L in the hot summer. These concentrations should be maintained throughout the aging process to limit *Brettanomyces* development (59) .

Brettanomyces can endure SO_2 treatment in barrels (45). The yeast survives treatment in areas of limited SO_2 contact: around bung holes, in the oak, and in the yeast sediment (lees). Work by Swaffield & Scott (61) showed yeast and bacteria could penetrate the porous cellular structure of oak barrels and establish active permanent sub-surface mixed cultures even after cleaning. Variations in wine composition (pH, anthocyanin concentration, nutrient content, and temperature) can affect SO_2 treatment in wine (60, 62) .

Yeast differ in sensitivity to SO_2 treatment in fruit juices and wines. *Brettanomyces* spp. can resist 500 mg/L SO_2 while *Pichia membranefaciens* & *Kloeckera apiculata* are less resistant (63) . In general, many yeast and bacteria are inhibited by 100 mg/L and less SO_2 (64) . Beech and Carr (65) found low concentrations of molecular SO_2 (0.625 mg/L) were toxic to species of the genera *Brettanomyces* and *Saccharomyces*.

Spread of *Brettanomyces*. An on-going debate exists in the wine industry as to the initial source and dispersion of *Brettanomyces* within a winery (45) . The surface of vineyard vegetation, flowers, fruit and soil are all possible sources of yeast flora, especially during the harvest season. Beech and Davenport (66) reviewed studies on the isolation of yeast populations -- from the previously named vineyard sources -- including *Saccharomyces, Hansenula, Pichia, Candida,* and *Kloeckera* within apple orchards. Yeast populations varied seasonally. Soil and vegetative surfaces on the orchard floor (i.e., clover & grasses) had the greatest cell counts in the autumn, as observed by Davenport; yeast counts were lower in the spring (66) .

Few conclusive studies have identified the source or spread of *Brettanomyces* within the vineyard except for use of contaminated equipment. Contaminated and improperly sanitized crush equipment, drains, barrels, transfer hoses, valves, pumps, and bottling equipment can all act as sources for further infection (33) . Harper (67) found the interior surface of plastic (polyvinyl chloride, polyethene, and plasticised nylon) pipes used commonly in the brewing industry maintained a variety of bacteria and yeast populations, including *Brettanomyces*.

Fruit flies and bees are involved in the spread of *Brettanomyces*. Yeast adhere to the body, legs, and wings of insects (68) . Van der Walt and van Kerken (33) recovered *Brettanomyces* from breeding and feeding areas of fruit flies (*Drosophilia*) within the winery. Yeast and other microorganisms are a normal part of the *Drosophilia* diet (69) . Using triturated *Drosophilia* under laboratory conditions, *Brettanomyces* were recovered externally from fruit flies 24 hrs after feeding on the yeast (33) .

Yeast survive internally in the nectar sac and intestinal tract of insects, particularly in pollinating bees (68, 70-72) . In India, Sandhu & Waraich (73) recovered 652 samples from pollinating bees, flower nectar, and fermented foods; all tested positive yielding 16 genera and 55 species of yeast. *Brettanomyces* were isolated from the honey stomachs of 8 of 271 pollinating bees and 5 of 137 fermented food samples. In Canada, Inglis (74) found *Brettanomyces* species associated with alfalfa leafcutter bees.

Other vectors may host *Brettanomyces* Yeasts associated with wine spoilage - - *Hansenula*, *Pichia*, *Candida*, and *Kloeckera* -- were isolated either internally or externally from earthworms and slugs found in soil (66) . Yeasts were also isolated from orchard insects: spiders (Arachnida sp.), St. Mark fever flies (Bilba sp.), bees, wasps, crane flies (*Ctenophora ornata*), aphids, manure flies (*Scotophaga sterioranum*), red & black froghoppers (*Cercopsis vulnerata*), ants, and fruit flies (*Drosophila* sp.). No mention of *Brettanomyces* isolation was made in this study.

Air is another source for *Brettanomyces* dispersion, although few articles have been written (66, 75) . In fruit orchards, Adams (76) exposed agar plates at ground level and isolated mainly molds (95%) and a few yeasts (5%) from fruit orchard air samples. Isolation of yeast colonies was hindered by early mold growth.   Adams identified 6 genera from 180 yeast isolates, the order in decreasing frequency: *Kloeckera*, *Cryptoccus*, *Torulopsis*, *Rhodotorula*, *Candida*, and *Brettanomyces* (66) .

Flavors Associated with *Brettanomyces* in Beer

"English character". Claussen (1) stressed "a general rule cannot be given for all cases, but the quality of *Brettanomyces* to be added must be regulated by local circumstances, more especially by the time the beer has to be stored and by the temperature of the storing room." A *Brettanomyces* inoculation with a wort of 1055 specific gravity and a room temperature of 24-27 °C would achieve the "English" character.

Schimwell confirmed these conditions: a 1.060 specific gravity was essential to achieve a "vinous" wine-like flavour (6) ; in contrast, a beer under 1.050 would produce an unpalatable and turbid beer with an objectionable, insipid flavor and aroma (77) . As Shimwell (6) noted, *Brettanomyces* can behave "as a desirable organism in one beer and an undesirable one at one and the same brewery".

Belgian beer. *Brettanomyces* species are essential in the production of characteristic fruity, ester-like aromas of spontaneous fermented Belgian beers (78, 79) : lambic, gueuze, kriek, and frambois. Others have described the Brett aroma in traditional Belgian beer as "smelling like horse sweat"; it is "the deliberate signature of the style" (80) .

According to Verachtert (81) , "lambic is the fermented wort and gueuze is derived from it after a secondary fermentation in the bottle... when cherries or extract are added during fermentation, the gueuze may be named 'kriek', and when raspberries are added it is called 'frambois' ". *Brettanomyces* species predominate in the latter part of the main fermentation and in bottles of lambic and gueuze beers.

Flavors Associated with *Brettanomyces* in Wine

"Mousiness". Heresztyn (44) first isolated and characterized the compounds and organisms responsible for "mousy taint" in Australian wines. Two isomers, 2-acetyl-1,4,5,6-tetrahydropyridine and 2-acetyl-3,4,5,6-tetrahydropyridine, produced by three

Brettanomyces yeast species and two *Lactobacillus* bacteria species were responsible for the mousiness in wines. The taint was produced by each class of microorganisms only in the presence of lysine and ethanol. The two 2-acetyltetrahydopyridine (ACTPY) isomers produced a "mousiness" described by Heresztyn as possessing an extremely unpleasant taste and odor; the odor was described as bready, cracker-like, and popcorn-like. Others have described the aroma as "a most disgusting smell reminiscent of mouse urine or acetamide"(82) .

Researchers at the Australian Wine Research Institute (83) have since identified two additional compounds associated with mousiness and produced by both *Brettanomyces* and *Lactobacillus* species: 2-acetyl-1-pyrroline (ACPY) and ethyltetrahydropyridine (ETPY). In tainted wines ETPY was found at concentrations below its high odor threshold, thereby contributing little to the sensory character of mousiness. ACTPY and ACPY were reported to be the main mousy taint compounds.

"Brettiness". "Volatile phenols can be considered as natural components in wines and beer, or as spoilage compounds when present in excessive amounts" (84) . *Brettanomyces* species can produce both mousiness and, at low concentrations of volatile phenolics, also "a distinct aroma described variously as cider-like, spicy, clove-like or phenolic... formed toward the end of fermentation" (84) . The ethyl phenols produced can exceed the sensory threshold 16-fold (85) , producing wines, at high concentrations of volatile phenolics, with distinct "barnyard", "stable", and "animal" phenolic odors (85-87) . Wines with high concentrations of phenolic odors are deemed "Bretty" by tasters.

As early as 1964 it was recognized that 4-ethyl phenol and 4-ethyl guaiacol were produced by yeast and bacteria during fermentation by the decarboxylation of the hydroxycinnamic acids p-coumaric and ferulic acid (88) . Later it was reported that among yeast only *Brettanomyces* species possess the metabolic ability to enzymatically decarboxylate hydroxycinnamic acids to produce ethyl derivatives (29, 89) . Heresztyn was the first to identify 4-ethyl phenol and 4-ethyl guaiacol as the major volatile phenolic compounds formed by *Brettanomyces* yeast (84) .

Lactic acid bacteria, including the typical "wine lactic acid bacteria" *Leuconostoc oenos* (85, 90) , can produce ethyl and vinyl derivatives by hydroxycinnamic acid metabolism (91) ; although, the minimal concentration produced in red wines by *Leuconostoc oenos* is insignificant compared to the odor threshold (85, 87) .

Descriptive and GCO Analysis – A Brief Synopsis

Cabernet wine comparison. One of the objectives of the study was to identify the odor-active compounds of wines with "Brett" flavor through sensory analysis and gas chromatography-olfactometry (GCO). Wines identified by their respective winemakers as having "Brett" character were evaluated by a trained expert sensory panel; also, using the technique CharmAnalysis (92-94) for GCO analysis, along with gas chromatography-mass spectrometry (GC-MS), odor-active compounds were identified by their respective Kovàts retention indices (95) . Contained below is a synopsis of our initial study; a detailed report of the sensory and GCO analyses will be given in subsequent publications.

Three wines from a single anonymous winery were chosen for our initial study. All three were non-sterile filtered Cabernet Sauvignon from the harvest years 1989, 1992, and 1994. According to the winemaker of these wines, the aroma of the 1989 Cabernet, identified as C89CS, was dominated by "Brett" character. The "Brett" aroma of the 1992 Cabernet (C92CS) was described as a mild aroma that contributed to the complexity of the wine. The 1994 Cabernet (C94CS) had no detectable "Brett" aroma. For comparitive ease, the C89CS had "high Brett", the C92CS "medium Brett", and the C94CS "no Brett" flavor.

4-ethyl phenol analysis was run at the winery after every quarterly racking for each wine; concentration values and *Brettanomyces* plate count numbers are included in table 2. All three wines contained viable populations of *Brettanomyces* yeasts capable of producing increased concentrations of ethyl derivatives from precursors during wine storage. At the time of our study, the "high Brett" bottled wine had 3.070 mg/L 4-ethyl phenol, the "medium Brett" had 1.736 mg/L, and the "no Brett" wine had 0.688 mg/L. In aqueous ethanol solution 4-ethyl phenol has a sensory threshold of 1.0 mg/L (97). Although it would appear that the "Bretty" wines were above threshold and the non-Bretty below, it is important to note that threshold values are specific to the test solution. Threshold values will vary depending on differences in ethanol concentration, temperature, and acidity in the wines (98).

Figure 1 is a spider plot of the perception of one of the sensory panelists in the comparison of the "high Brett", "medium Brett", and "no Brett" wines. In generating the descriptors, panelists agreed to include groups of descriptors under a general descriptor name; for example, the descriptor 'plastic' included typical "Brett" descriptors such as horse sweat, rubber hose, and band aid. In general, fruity, floral, spicy, earthy, woody aromas predominated while no 'plastic' aromas were detected in the "no Brett" wine. The opposite effect is evident in the "high Brett" wine, while it would appear that the "medium Brett" wine was somewhere in-between.

Figure 2 details a GCO chromatogram for the "high Brett" wine. The 15 most odor active compounds were identified by GCO and GC-MS. The odor active compounds were ranked from 1-15, 1 being the most odor active compound. The numbers above each peak in the chromatogram correspond to the compound and associated descriptor in the table.

Isovaleric acid (3-methyl butanoic acid) was found to be the dominant odorant in the "high Brett" wine as detected by CharmAnalysis. The odor described by the GCO sniffer was 'rancid'; the chemical identity of the odorant was confirmed by GC-MS. This acid is produced in wine by yeast as a metabolic byproduct of protein (99). Volatile phenolic compounds, such as 4-ethyl guaiacol, guaiacol, and 4-ethyl phenol, were also among the dominate odor active compounds in this wine; however, the individual contribution by each of the three phenolics was half or less than the odor activity of isovaleric acid.

The chemical identity of the second most dominate odorant in the "high Brett" wine, identified as 'plastic' by CharmAnalysis, remains 'unknown'. This compound had a Kovàts retention index at 1434 on an OV-101 column. The EI-MS of this unknown includes a base peak at m/z {%} 107 {100}, a molecular ion at m/z 138 {28}, and fragment ions at m/z 139 {1}, 108 {8}, 91 {3}, 79 {5}, 78 {6}, 77 {20}, 63

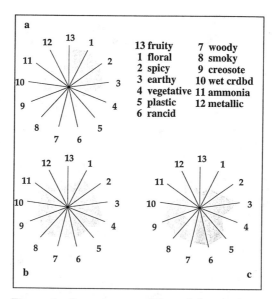

Figure 1. Sensory perception of three wines:
(a) C94CS "no Brett", (b) C92CS "medium Brett", (c) C89CS "high Brett"

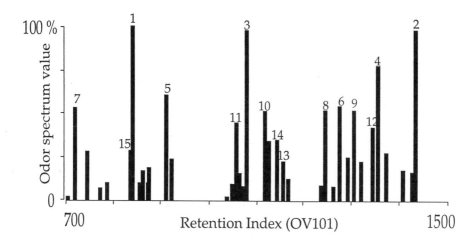

	Odor active compound	Odor descriptor	Retention index	Charm value	Spectral value
1	isovaleric acid	rancid	842	33635	100
2	unknown	plastic	1434	32089	98
3	2-phenyl ethanol	floral	1080	31762	97
4	ß-damascenone	fruity	1357	19808	77
5	unknown	vitamin tablets	912	11824	60
6	unknown	vegetative	1276	9553	53
7	isoamyl alcohol	fruity	718	9067	52
8	4-ethyl guaiacol	spicy	1246	8800	51
9	ethyl decanoate	plastic	1307	8609	51
10	cis-2-nonenal	burning tires	1118	8285	50
11	guaiacol	plastic	1058	6092	43
12	unknown	plastic	1345	5677	41
13	4-ethyl phenol	plastic	1145	3823	34
14	trans-2-nonenal	burning tires	1126	3478	32
15	ethyl-2-methyl butyrate, ethyl-3-methyl butyrate	fruity	836	2441	27

Figure 2. Odor spectrum gas chromatogram of the "high Brett" wine. GC/MS & GCO identification on an OV101 column of the 15 most potent odor potent compounds.

{4}, 51 {6}, and 39 {6}. Together the GCO and mass spectral data closely match a number of phenolic compounds. The identity remains unknown until the odor and retention index can be matched by a chemical standard. Chemical identification of GCO odorants relies on the confirmation with authentic chemical standard. The general rule applies: "if two chemicals have the same retention time and the same odor character, they are the same chemical. If one of the chemicals is an authentic standard, the other has been identified with considerable certainty (93) ".

Comparing the odor spectrum gas chromatograms of the three wines, in Figure 3, a general effect was observed. 'Floral', 'fruity' compounds were the dominant odorants in the "no Brett" wine while 'rancid', 'plastic' odors accounted for 1/3 or less of the odor activity; in the "medium Brett" wine, the 'floral', 'fruity' compounds decrease to 1/2 or less of the odor activity while the 'rancid', 'plastic' compounds increase to 2/3; in the "high Brett" wine, the 'rancid', 'plastic' compounds were the dominant odorants while the 'floral', 'fruity' compounds were far less dominant. The 'floral' odorant identified as 2-phenyl ethanol was the dominant compound in "no Brett" and "medium Brett" wines; in the "high Brett" wine, it was equally as dominant as isovaleric acid and the unknown compound. The 'fruity' odorant ß-damascenone was equally dominant among the three wines; for this reason, it should not be considered as a contributor to "Brett" aroma.

Brett flavor in wine? The question still remains: what is "Brett" flavor? Results from our initial work indicates that "Brett" aroma in wine is a complex mixture of odor-active compounds, including acids, alcohols, aldehydes, ketones, esters, and phenolics. Analysis by gas chromatography-olfactometry revealed two predominate odor-active compounds responsible for the Brett flavor in the wines studied: isovaleric acid and a second unknown compound; other identified odor-active compounds included 2-phenyl ethanol, isoamyl alcohol, cis-2-nonenal, trans-2-nonenal, ß-damascenone, ethyl decanoate, guaiacol, 4-ethyl guaiacol, 4-ethyl phenol. Our findings are a snapshot into the much larger picture know as Brett flavor. Ultimately this preliminary investigation requires the descriptive analyses of many more wines to know what odor active compounds describe the flavor know as "Brett".

Acknowledgments Special thanks are due to Ed Lavin and Peter Ong for their enthusiastic interest and collaboration in the quest to identify the odor active compounds by GCO and GC/MS; and to Jeanne Samimy of the Cornell NYSAES Geneva library for always making our literature searches easier.

Literature Cited
1. Claussen, N. H. *J. Inst. Brew.* **1904,** *10*, pp. 308-331.
2. Riesen, R. In *Undesirable fermentation aromas*; Henick-Kling, T., Ed.; NYSAES; Wine Aroma Defects Workshop; Geneva, NY; 1992; pp. 1-43.
3. Claussen, N. H. *Improvements in and connected with the manufacture of English beers or malt liquors and in the production of pure yeast cultures for use therein*; Eng. Pat., 28,184; Dec. 22, 1903; p. 204.
4. Parsons, C. M. In *Alcoholic Beverages*; Birch, G. G.; Lindley, M. G., Eds.; Elsevier Applied Science Publishers: London, 1985, pp. 43-50.

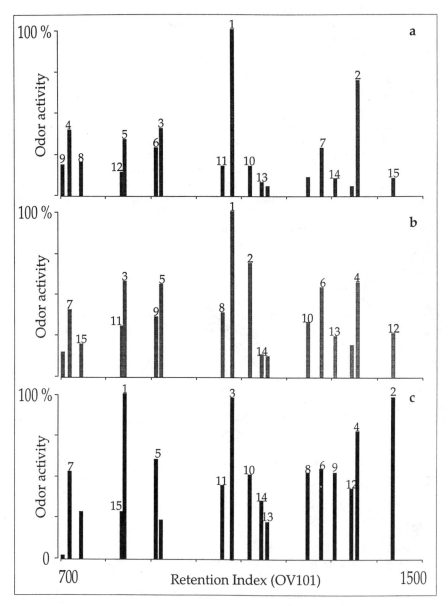

Figure 3. Odor spectrum gas chromatograms on an OV101 column - the top 15 odor active compounds: (a) "no Brett", (b) "medium Brett", (c) "high Brett". Numbers on the chromatogram refer to the chemical structures in figure 4.

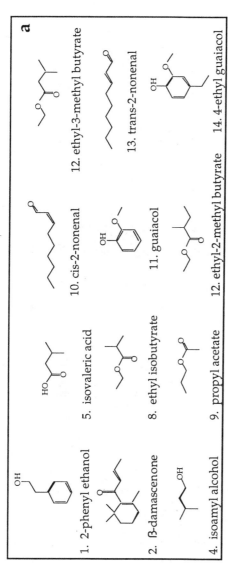

Figure 4. Chemical structures of the top 15 odor active compounds identified by CharmAnalysis and GC/MS analysis in three Cabernet Sauvignon wines: (a) C94CS, "no Brett", (b) C92CS, "medium Brett", (c) C89CS, "high Brett". Numbers correspond to odor activity, 1 being the most odor active compound. Unidentified compounds are (a) 3,6,7,15, (b) 5,6,9,12, and (c) 2,5,6,12.

Continued on next page.

110

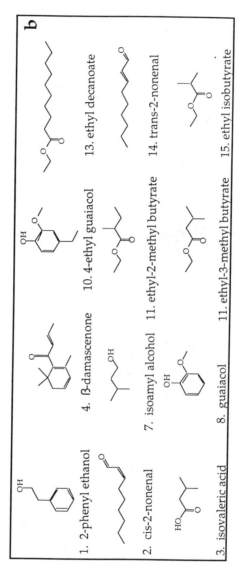

Figure 4. *Continued.*

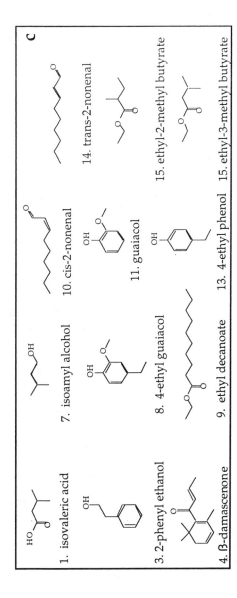

C

1. isovaleric acid

3. 2-phenyl ethanol

4. ß-damascenone

7. isoamyl alcohol

8. 4-ethyl guaiacol

9. ethyl decanoate

10. cis-2-nonenal

11. guaiacol

13. 4-ethyl phenol

14. trans-2-nonenal

15. ethyl-2-methyl butyrate

15. ethyl-3-methyl butyrate

Figure 4. *Continued.*

5. Gilliland, R. B. *J. Inst. Brew.* **1961**, *67*, pp. 257-261.
6. Shimwell, J. L. *Amer. Brewer* **1947**, *80*, pp. 21-22, 56-57.
7. Anonymous *Brewers J.* **1959**, *95*, pp. 41-42.
8. Custers, M. T. J. *Onderzeokingen over het Gistseglacht Brettanomyces*; thesis, Delft University, 1940.
9. Kufferath, H.; van Lear, M. H. *Bulletin de la Societé Chimique de Belgique* **1921**, *30*, p. 270.
10. Krumbholz, G.; Tauschanoff, W. *Zentralblatt füer Bakteriologie* **1933**, *88*, pp. 366-373.
11. Wikén, T.; Scheffers, W. A.; Verhaar, A. J. M. *Antonie van Leeuwenhoek* **1961**, *27*, pp. 401-433.
12. Scheffers, W. A. *Nature* **1966**, *210*, pp. 533-534.
13. Burk, D. In *A colloquial consideration of the Pasteur and neo-Pasteur effects*; Ponder, E., Ed.; The Darwin Press, New Bedford, MA; Cold Spring Harbor Symposia on Quantitative Biology; Cold Spring Harbor, NY; 1939; pp. 420-459.
14. Scheffers, W. A. *Antonie van Leeuwenhoek* **1979**, *45*, p. 150.
15. Carrascosa, J. M.; Dolores Viguere, M.; Nunez de Castro, I.; Scheffers, W. A. *Antonie van Leeuwenhoek* **1981**, *47*, pp. 209-215.
16. Wijsman, M. R.; van Dijken, J. P.; Scheffers, W. A. *Antonie van Leeuwenhoek* **1984**, *50*, p. 112.
17. Osterwalder, A. *Zentralblatt füer Bakteriologie II* **1912**, *33*, pp. 257-272.
18. Schanderl, H. *Die Mikrobiologie des Weines*; Verlagsbuchhandlung Eugen Ulmer: Stuttgart, 1950.
19. Schanderl, H.; Draczynski, M. *Wein und Rebe* **1952**, *20*, p. 462.
20. van der Walt, J. P.; van Kerken, A. E. *Antonie van Leeuwenhoek* **1958**, *25*, pp. 145-151.
21. van der Walt, J. P.; van Kerken, A. E. *Antonie van Leeuwenhoek* **1958**, *24*, pp. 241-252.
22. Florenzano, G. *Ricerca. Scient.* **1950**, *20*, p. 1494.
23. Florenzano, G. *Atti. Accad. Ital. vite. vino* **1951**, *3*.
24. Barret, A.; Biden, P.; Andre, L. *C. R. Acad. Agric.* **1955**, *41*, p. 426.
25. Johnson, H. *The World Atlas of Wine*, IV ed.; Simon & Schuster: New York, 1994, p. 137.
26. Galzy, P.; Rioux, J. A. *Progr. Agric. Vitic.* **1955**, *144*, pp. 365 - 370.
27. Domercq, S. *Etude et classification des levures de vin de la Gironde*; thesis, Bordeaux, 1956.
28. Peynaud, E.; Domercq, S. *Arch. Mikrobiol.* **1956**, *24*, pp. 266-280.
29. Chatonnet, P.; Dubourdieu, D.; Boidron, J.-N.; Pons, M. *J. Sci. Food Agric* **1992**, *60*, pp. 165-178.
30. Larue, F.; Rozes, N.; Froudiere, I.; Couty, C.; Perreira, G. P. *J. Int. Sci. Vig. Vin* **1991**, *25*, pp. 149-165.
31. van der Walt, J. P.; van Kerken, A. E. *Antonie van Leeuwenhoek* **1959**, *25*, pp. 449-457.
32. van der Walt, J. P.; van Kerken, A. E. *Antonie van Leeuwenhoek* **1960**, *26*.
33. van der Walt, J. P.; van Kerken, A. E. *Antonie van Leeuwenhoek* **1961**, *27*, pp. 81-90.

34. Zyl, J. A. v. *Turbidity in South African dry wines caused by the development of the Brettanomyces yeasts*, Science Bulletin No. 381, Viticultural and Enological Research Institute, Stellenbosch, South Africa.

35. Verona, O. *Ann. Fac. Agr. Pisa* **1951**, *12*, pp. 123-145.

36. Zardetto de Toledo, O.; Teixeira, C. G.; Verona, O. *Ann. Micriobiol. Enziol.* **1959**, *9*, pp. 22-34.

37. Mavlani, M. I. *Antonie van Leeuwenhoek* **1968**, *35*, p. D3.

38. Ibeas, J. I.; Lozano, I.; Perdigones, F.; Jimenez, J. *Appl. Environ. Microbiol.* **1996**, *62*, pp. 989-1003.

39. Arjun, V. M. In *Biology and activities of yeasts*; Skinner, F. A.;Passmore, S. M.; Davenport, R. R., Eds.; Academic Press Inc.: London, 1979, p. 301.

40. Mateo, J. J.; Jimenez, M.; Huerta, T.; Pastor, A. *International Journal of Food Microbiology* **1991**, *14*, pp. 153-160.

41. Henschke, P. *Technical Review,*, Australian Wine Research Institute, December 1996, *105*, pp. 12-13.

42. Wright, J. M.; Parle, J. N. *New Zealand Journal of Agricultural Research* **1974**, *17*, pp. 273-278.

43. Tucknott, O. G.; Davies, P. A.; Rosser, P. R. *Production of mousy taints by microorganisms*, Annual, University of Bristol, Long Ashton Research Station, 1976, p. 140.

44. Heresztyn, T. *Am. J. Enol. Vitic.* **1986**, *37*, pp. 127-132.

45. Hock, S. In *Practical Winery & Vineyard*, 1990; Vol. 10, pp. 26-31.

46. Vilas, M. In *Vineyard & Winery Management*, 1993, Vol. 19, pp. 33-35.

47. Peynaud, E. *Knowing and making wine*; John Wiley & Sons: New York, 1984, p. 102.

48. Trioli, G. In *Survey of studies in Italy on management of factors affecting alcoholic fermentations*; ; ; Wine Spoilage Microbiology Conference; Fresno, CA; 1996; pp. 42-43.

49. Querol, A.; Jiménez, M.; Huerta, T. *J. Fd Sci.* **1990**, *55*, pp. 1603-1606.

50. Beech, F. W.; Carr, J. G. *J. Gen. Microbiol.* **1955**, *12*, pp. 85-94.

51. Kunkee, R. E.; Amerine, M. A. In *The Yeasts*; Rose, A. H.; Harrison, J. S., Eds.; Academic Press: London, 1970; Vol. III, pp. 73-146.

52. Beech, F. W.; Davenport, R. R.; Mossel, D. A. A.; Kijkmann, K. E.; Koopmans, M.; de Jong, J.; Put, H. M. C.; Tilbury, R. H. In *Media and methods for growing yeasts: proceedings of a discussion meeting*; Skinner, F. A.;Passmore, S. M.; Davenport, R. R., Eds.; Academic Press; Biology and Activities of Yeasts; 1980; pp. 260-277.

53. Phaff, H. J.; Starmer, W. T. In *Specificity of natural habitats for yeasts and yeast-like organisms*; Skinner, F. A.;Passmore, S. M.; Davenport, R. R., Eds.; Academic Press; Biology and Activities of Yeasts; 1980; pp. 79-100.

54. Blazer, R. M.; Schleußner, C. D. *46th Annual Meeting, American Society for Enology and Viticulture* **1995**.

55. Fugelsang, K. C.; Osborn, M. M.; Muller, C. J. In *Beer and Wine Production - Analysis, Characterization, and Technical Advances*; Gump, B. H.; Pruett, D. J., Eds.; American Chemical Society: Washington, DC., 1993; Vol. ACS Symposium Series 536, pp. 110-131.

56. Boulton, R. B.; Singleton, V. L.; Bisson, L. F.; Kunkee, R. E. *Principles and Practices of Winemaking*, 1st ed.; Chapman & Hall: New York, 1996, p. 604

57. Heimoff, S. In *Wine Business Monthly*, 1996; Vol. 3, pp. 37-40.

58. Frey, S.; Henry, T.; Mahaney, P.; Paris, P. *Am. J. Enol. Vitic.* **1996**, *47*, p. 348.

59. Chatonnet, P.; Boidron, J. N.; Dubourdieu, D. *J. Int. Sci. Vig. Vin* **1993**, *27*, p. 277.

60. Smith, C. R. *Studies of sulfur dioxide toxicity for two wine yeasts*; personal communication, University of California, Davis, 1996.

61. Swaffield, C. H.; Scott, J. A. *J. Am. Soc. Brew. Chem.* **1995**, *53*, pp. 117-120.

62. Rose, A. H. In *The Yeasts*; Rose, A. H.; Harrison, J. S., Eds.; Academic Press Inc.: London, 1987; Vol. 2, pp. 21-32.

63. Reed, G.; Peppler, H. J. *Yeast Technology*; AVI Publishing Co.: Westport, Connecticut, 1973, p. 366.

64. Hammond, S. M.; Carr, J. G. *Symp. Ser. Soc. Appli. Bacter.* **1976**, *5*, pp. 89-110.

65. Beech, F. W.; Burroughs, L. F.; Timberlake, C. F.; Whiting, G. C. *Bull. O.I.V.* **1979**, *586*, pp. 1001-1022.

66. Beech, F. W.; Davenport, R. R. In *The Yeasts*; Rose, A. H.; Harrison, J. S., Eds.; Academic Press: London, 1970; Vol. III, pp. 73-146.

67. Harper, D. R. *Process Biochem.* **1980/81**, *16*, pp. 2-7.

68. Phaff, H. J.; Miller, M. W.; Mrak, E. M. *The Life of Yeasts*, 2nd ed.; Harvard University Press: Cambridge, MA, 1978, p. 341.

69. Walker, H. W.; Ayres, J. C. In *The Yeasts*; Rose, A. H.; Harrison, J. S., Eds.; Academic Press: London, 1970; Vol. 3, pp. 464-527.

70. Griliione, P.; Federici, F.; Miller, M. W. In *Yeasts from honey bees (Apis mellifera L.)*; Fifth International Yeast Symposium; London, Canada; 1980.

71. Batra, L. R.; Millner, P. H. *Mycol.* **1974**, *66*, pp. 942-950.

72. El-Tabey, A. M.; Shiata, A.; Mrak, E. M. *Am. Natural.* **1951**, *85*, pp. 381-383.

73. Sandu, D. K.; Waraich, M. K. In *Ecology of yeasts associated with pollinating bees, nectary glands of flowers and fermented foods*; Fifth International Yeast Symposium; London, Canada; 1980; pp. 535-539.

74. Inglis, G. D.; Sigler, L. *Mycol.* **1992**, *84*, pp. 555-570.

75. Beech, F. W. In *The Yeasts*; Rose, A. H.; Harrison, J. S., Eds.; Academic Press: London, 1993; Vol. V, pp. 169-213.

76. Adams, A. M. *Can. J. Microb.* **1964**, *10*, pp. 641-646.

77. Shimwell, J. L. *J. Inst. Brew.* **1938**, pp. 563-572.

78. Spalpen, M.; van Oevelen, D.; Verachtert, H. *J. Inst. Brew.* **1978**, *84*. pp. 278-282

79. Guinard, J. X.- *Lambic*, Classic Beer Style Series 3, Brewers Publications Book, Boulder, Colorado, pp. 25-32.

80. Bergen, R. *Brewing Techniques* September/October, **1993.**

81. Verachtert, H.; Shanta Kumara, H. M. C.; Dawoud, E. In *Yeast biotechnology and biocatalysis*; Verachtert, H.; de Mot, R., Eds.; Marcel Dekker, Inc.: New York, 1990; Vol. 5, pp. 429-478.

82. Sponholz, W.-R. In *Wine Microbiology and Biotechnology*; Fleet, G. H., Ed.; Harwood Academic Publishers, 1993, pp. 395-420.

83. Henschke, P. *Microbiology Group Report*, 42nd Annual Report, The Australian Wine Research Institute, 30 June 1996.

84. Heresztyn, T. *Archives of Microbiology* **1986,** *146*, pp. 96-98.

85. Chatonnet, P.; Dubourdieu, D.; Boidron, J. N. *Am. J. Enol. Vitic.* **1995,** *46*, pp. 463-468.

86. Etiévant, P. X.; Issanchou, S.; Marie, S.; Ducruent, V.; Flanzy, C. *Sci. Aliment* **1989,** *9*, pp. 19-33.

87. Chatonnet, P.; Boidron, J. N.; Pons, M. *Sci. Aliment* **1990,** *10*, pp. 565-587.

88. Steinke, R. D.; Paulson, M. C. *J. Agric. Food Chem.* **1964,** *12*, pp. 381-387.

89. Edlin, D. A. N.; Narbad, A.; Dickinson, J. R.; Lloyd, D. *FEMS Microbiology Letters* **1995,** *125*, pp. 311-316.

90. Cavin, J. F.; Andioc, V.; Etiévant, P. X.; Divies, C. *Am. J. Enol. Vitic.* **1993,** *44*, pp. 76-80.

91. Lindsay, R. F.; Priest, F. G. *J. Appl. Bact.* **1975,** *39*, pp. 181-187.

92. Acree, T. E.; Barnard, J.; Cunningham, D. G. *Food Chem.* **1984,** *14*, pp. 273-286.

93. Acree, T. E.; Bernard, J. In *Trends in Flavor Research*; Maarse, H.; van der Heij, D.G., Eds.; Elsevier Science: New York, 1994, pp. 211-220.

94. Acree, T. E. *Analytical Chemistry: News & Features* **1997,** *ACS*, pp. 170A-175A.

95. Kovàts, E. *Advan. Chromatogr.* **1965,** *1*, pp. 229-247.

96. Mitrakul, C. *Evaluation of DNA fingerprinting methods for identification of Brettanomyces/Dekkera yeasts from wines.* thesis, Cornell University, 1997.

97. Williams, A. A. *J. Inst. Brew.* **1974,** *80*, pp. 455-470.

98. Noble, A. C. In *Wine Analysis,* Linskens, H. F.; Jackson, J. F., Eds.; Springer-Verlag: Heidelberg, 1988, pp. 9-28.

99. Etiévant, P. X. In *Volatile Compounds in Foods and Beverages*, Maarse, H., Ed.; Marcel Dekkar, Inc.: New York, 1991; Vol. 1st ed., pp. 483-546.

Chapter 9

Rationalizing the Origin of Solerone (5-Oxo-4-hexanolide)

Biomimetic Synthesis, Identification, and Enantiomeric Distribution of Key Metabolites in Sherry Wine

D. Häring, B. Boss, M. Herderich, and P. Schreier

Food Chemistry, University of Würzburg, Am Hubland, D-97074 Würzburg, Germany

A biomimetic synthesis of solerone (5-oxo-4-hexanolide) **1** using both enzymatic and acid-catalyzed reactions was performed. Starting from L-glutamic acid 5-ethyl ester **2** enzymatic oxidative deamination followed by subsequent decarboxylation of the corresponding 2-oxoglutaric acid 5-ethyl ester **3** led to ethyl 4-oxobutanoate **4**. In the presence of pyruvate, **4** served as key substrate for a novel acyloin condensation catalyzed by pyruvate decarboxylase [EC 4.1.1.1] from *Saccharomyces cerevisiae*. Finally, the resulting ethyl 4-hydroxy-5-oxo-hexanoate **5** was easily converted into **1** in the presence of acid. The acyloin condensation of **3** with acetaldehyde to ethyl 5-hydroxy-4-oxohexanoate **6** revealed an alternative route to **1**. Confirming the relevance of the proposed biogenetic pathway, all solerone precursors were identified in sherry by GC-MS analysis. Additionally, the enantiomeric distribution of the chiral progenitors and solerol (5-hydroxy-4-hexanolide) in sherry wines was determined by multidimensional gas chromatography-mass spectroscopy (MDGC-MS).

Solerone (5-oxo-4-hexanolide) **1** is a known constituent of wine (*1*), in particular, flor sherry (*2*). Recently, it has also been identified in dried figs (*3*). While the contribution of solerone **1** to the aroma of sherry is controversially discussed (*4*), it is generally accepted that **1** is enzymatically formed during the course of sherry fermentation under oxidative conditions. However, experimental information on both enzymes and key metabolites involved in the biosynthesis of **1** is rather scarce to date. Previous ^{14}C-labeling experiments have indicated the involvement of glutamic acid and ethyl 4-oxobutanoate **4** (*5,6*), but **4** has not been detected in sherry as yet. To rationalize the origin of solerone **1** by biomimetic synthesis and identification of the metabolites involved, both enzymatic and acid-catalyzed reactions were performed. In addition, enantiomeric distribution of chiral solerone progenitors was evaluated by multidimensional gas chromatography-mass spectroscopy (MDGC-MS). The results are described in this paper.

Results and Discussion

Chemical Synthesis of Reference Compounds. Prerequisite for the unequivocal identification of compounds in microscale analysis is the availability of authentic chromatographic and spectroscopic data. The linear "ex chiral pool" synthesis of (S)-solerone **1** (7) started from enantiomerically pure L-glutamic acid and yielded an enantiomeric excess (e.e.) of 90 % (S)-**1**. The related (S)-ethyl 4-hydroxy-5-oxohexanoate **5** (e.e. 89 %) was synthesized by mild ethanolysis of (S)-**1** for the first time. Accordingly, a chemical synthesis of ethyl 5-hydroxy-4-oxohexanoate **6** had to be developed. Starting from 2-acetylfuran racemic 4-oxo-5-hexanolide **7** was formed in a six step synthesis (8). δ-Lactone **7** was converted into its hydroxy ester **6** via acid-catalyzed ethanolysis. All compounds were identified by ^{1}H- and ^{13}C-NMR spectroscopy as well as by their EI mass spectra (9). The stereochemical analysis of lactones **1** and **7** was performed by multidimensional gas chromatography-mass spectrometry (MDGC-MS) with chiral cyclodextrin phases. Enantioseparation of reference α-ketols **5** and **6** was achieved by GC-MS on a chiral stationary phase (chiral GC-MS) (10).

Biomimetic Synthesis of Solerone. We applied pyruvate decarboxylase [EC 4.1.1.1] (PDC) as key enzyme for the biomimetic synthesis elucidating the formation of solerone **1** (Figure 1). The thiamine diphosphate depending enzyme from *Saccharomyces cerevisiae* is responsible for the decarboxylation of pyruvate in the course of alcoholic fermentation. After loss of carbon dioxide from 2-oxoacids the resulting aldehyde is released. Alternatively, the cofactor-bound decarboxylation product can react with a further aldehyde. By the latter acyloin condensation a new carbon-carbon bond will be formed, thus opening a biosynthetic way to α-hydroxy carbonyl compounds (11,12).

While in the presence of 2-oxoglutaric acid neither decarboxylation nor acyloin condensation had been observed, as expected from previously published results (13), we succeeded in the enzymatic conversion of the mono ethyl ester **3** to ethyl 4-oxobutanoate **4**, using both whole yeast cells (*Saccharomyces cerevisiae*) and purified PDC. The oxo ester **4** served as substrate for a second reaction catalyzed by PDC. Formation of a new carbon-carbon bond was accomplished in the presence of pyruvic acid which acted as donor of a C_2-unit. Thus, ethyl 4-hydroxy-5-oxohexanoate **5** was obtained for the first time as the result of an enzymatic acyloin condensation. Finally, traces of acid induced the lactonization of hydroxyester **5**, indicating it as direct precursor of solerone **1** (Figure 1).

These results demonstrated the importance of the ethyl ester function, which has not found attention in previous discussions on the biogenesis of **1** (14,15). Consequently, we focused our interest on glutamic acid 5-ethyl ester **2** as potential precursor of the corresponding 2-oxoglutaric acid 5-ethyl ester **3**. Our view was supported by earlier ^{14}C-labeling experiments, in which the involvement of L-glutamic acid in the biosynthesis of **1** has been suggested (5,6). In addition, diethyl glutamate has already been identified in sherry (16,17). While amino acids can be transformed to 2-oxoacids by pyridoxal depending transaminases as well, we applied the oxidative deamination of **2** catalyzed by L-amino acid oxidase [EC 1.4.3.2] (18). The use of an oxygen electrode enabled direct monitoring of the reaction. Hydrogen peroxide had to be destroyed with

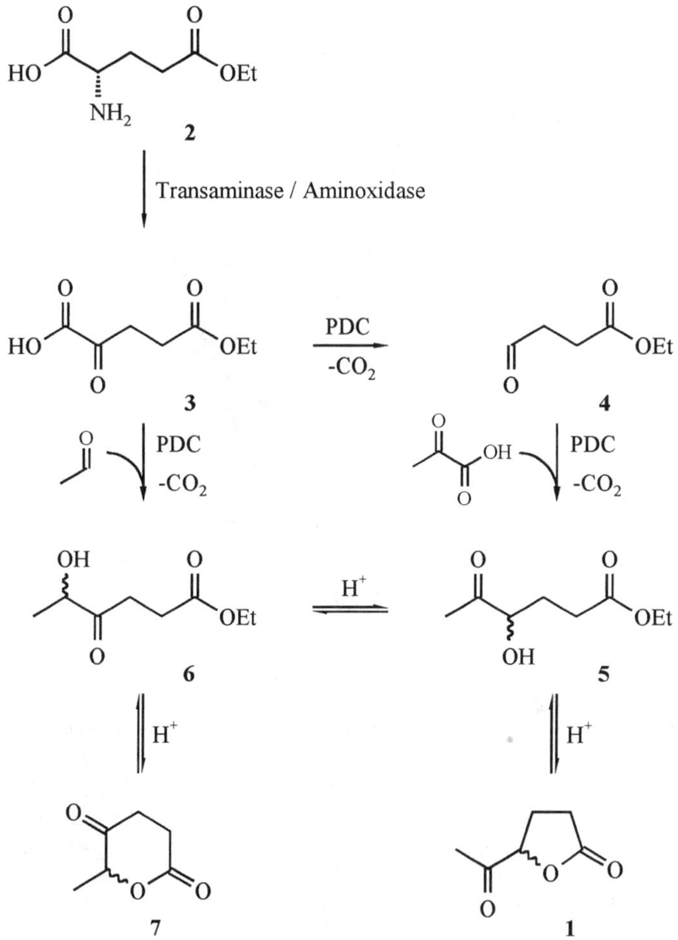

Figure 1. Postulated biogenesis of solerone **1** and related sherry constituents.

catalase [EC 1.11.1.6] in order to avoid oxidative decarboxylation and degradation of **3**. By this oxidative enzymatic reaction 5-ethyl 2-oxoglutarate **3** was obtained with excellent yield.

The proposed biosynthetic pathway describing the transformation of ethyl glutamate **2** to solerone **1** *via* ethyl 4-oxobutanoate **4** is in good agreement with previously reported radiotracer experiments (*5,6,19*). In addition, we evaluated another yet unknown route to solerone **1**. Starting from 2-oxoglutaric acid 5-ethyl ester **3** the PDC-catalyzed acyloin condensation with acetaldehyde yielded in one step ethyl 5-hydroxy-4-oxohexanoate **6**. Acid-catalyzed lactonization of **6** led to 4-oxo-5-hexanolide **7**. As by-products substantial amounts (up to 40 %) of ethyl 4-hydroxy-5-oxohexanoate **5** and solerone **1** were formed. A similar rearrangement yielding solerol (5-hydroxy-4-hexanolide) has been observed on storage of 4-hydroxy-5-hexanolide (*20*).

Identification of the Key Metabolites 2-7 in Sherry Wine. In order to identify the postulated metabolites in sherry wine, a Manzanilla and an Oloroso sherry were analyzed. Neutral and acidic sherry constituents were analyzed by GC-MS after solvent extraction. Selective extraction of amino acid ethyl ester **2** was achieved using a strongly acidic cation exchange resin. To avoid the artefactual formation of ethyl esters the method described by Herraiz and Ough (*17*) was chosen. Control experiments revealed no amino acid ethyl ester formation during extraction. Prior to GC-MS analysis mono ethyl ester **2** had to be derivatized with trifluoroacetic anhydride yielding ethyl 3-(2'-trifluoro-methyl-5'(4'*H*)-oxazolon-4'-yl)-propanoate (*21*). Diethyl glutamate was simultaneously detected as N-trifluoroacetyl derivate. For the first time, L-glutamic acid 5-ethyl ester **2**, 2-oxoglutaric acid 5-ethyl ester **3**, ethyl 4-oxobutanoate **4**, ethyl 4-hydroxy-5-oxohexanoate **5**, ethyl 5-hydroxy-4-oxohexanoate **6** and 4-oxo-5-hexanolide **7** were identified in sherry by comparison of chromatographic and spectroscopic data with those of authentic reference substances. The newly identified compounds **2-7** are listed in Table I.

Table I. Solerone and key progenitors as identified in sherry samples by GC-MS

Compound	Ri	Concentration[a]
L-Glutamic acid 5-ethyl ester **2**	1983[b]	+
2-Oxoglutaric acid 5-ethyl ester **3**	2374	+++
Ethyl 4-oxobutanoate **4**	2280	+
Ethyl 4-hydroxy-5-oxohexanoate **5**	2135	+
Ethyl 5-hydroxy-4-oxohexanoate **6**	2189	+
4-Oxo-5-hexanolide **7**	2194	+
Solerone **1**	2113	+

[a] Relative data evaluated by addition of external standard (2.0 mg/L 2-undecanol; Ri = 1704) + < 0.5 mg/L; ++ 0.5-10 mg/L; +++ 10-50 mg/L. The concentrations were comparable in both sherry wines.

[b] Ri of the trifluoroacetyl derivate.

Stereochemical Analysis of Solerone Progenitors. In order to determine enantiomeric distributions of chiral constituents from complex matrices such as food the use of multidimensional HRGC techniques (MDGC) has been recommended due to their selectivity and sensivity as well as the simplified cleanup procedures (22-24). Coupling of the achiral precolumn to the chiral main column was realized by the "moving column stream switching" (MCSS) system (25,26) in which columns are connected utilizing a dome shaped glass tip. The cut is performed by moving the outlet of the precolumn close to the inlet of the main column. As a result, surface contact of analytes was minimized.

After establishing the MDGC-MS method applying the MCSS system we determined the enantiomeric ratios of both α-ketols **5** and **6** as obtained by enzyme-catalyzed biomimetic synthesis (9). The acyloin condensation of ethyl 4-oxobutanoate **4** in presence of pyruvate was catalyzed by purified pyruvate decarboxylase (from *Saccharomyces carlsbergensis*) or intact cells of *S. cerevisiae* and yielded (R)-**5** (e.e. = 78 %). The same biotransformation was applied to 5-ethyl glutarate **3** and acetaldehyde yielding the first eluting enantiomer of **6** (e.e. = 75 %) with unknown absolute configuration (10).

Finally, α-ketols **5** and **6** as well as the related lactones solerone **1**, 4-oxo-5-hexanolide **7** and solerol **8** were analyzed by means of MDGC-MS in two sherry extracts (Table II) (10). The enantiomeric ratios of solerone **1** and solerol diastereomers **8a/b** are in good agreement with previously published data (27,28). Solerol **8** was detected in hundredfold amounts compared to **1,4-7** and seems to represent the end of the biosynthetic pathway leading to sherry lactones. GC-MS analysis revealed comparable amounts of the diastereomers **8a** and **8b**. But in contrast to **1**, which has been demonstrated to undergo racemisation on storage (4,27), configuration of the major lactones **8a/b** could be utilized as probe for demonstrating the relevance of the proposed biosynthetic pathway (9). Clearly, the (4R)-configurated solerol isomers dominated the (4S)-isomers with a ratio of 3 : 2 (Table II, Figure 2), thus demonstrating the relevance of α-ketol (R)-**5** as obtained from biomimetic synthesis. In addition, reduction of the oxo-function in **1** yielded (5R)-solerol with even higher enantioselectivity.

Table II. Enantiomeric distribution of sherry constituents as determined by MDGC

Compounds	Enriched Enantiomer	Enantiomeric Excess [%]	
		Manzanilla	*Oloroso*
1	S	16	0
5	S	20	54
6	– a	64	70
7	– a	38	52
8a	(4S, 5R)	22	58
8b	(4R, 5R)	99	99

a Assignment of the enantiomers unknown.

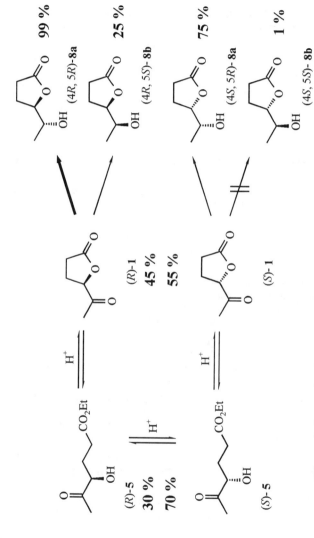

Figure 2. Stereochemical details of the biogenesis of solerone **1** and solerol **8**. The average enantiomeric distribution of **1**, **5** and **8** in sherry considers literature data (*26,28*) and own results.

While α-ketol **6**, serving as precursor for 4-oxo-5-hexanolide **7** and solerone **1**, was demonstrated to occur in the sherry samples with an enantiomeric ratio comparable to that as obtained from biomimetic synthesis, in the case of solerone **1** and its direct progenitor **5** the situation is much more complex. First, one has to consider the equilibrium between lactone **1** and α-ketols **5** and **6** which can rationalize the slow racemisation of **1** as observed previously (*4,27*). Second, as demonstrated by the high concentration of (4*R*)-solerol, it is apparent that mainly (4*R*)-solerone underwent reduction. Considering the equilibrium situation between **1** and **5** as discussed above, one could expect that not only (*R*)-lactone **1** but also its direct precursor (*R*)-**5** should be depleted. As result of both equilibria and decline of (*R*)-**5**, the actual excess of α-ketol (*R*)-**5**, as formed by PDC from *S. cerevisiae*, might have been decreased in favor of (*S*)-**5**. However, one should take into consideration that thiamin diphosphate depending enzymes other than the utilized decarboxylase from *S. cerevisiae* could be involved in the biosynthesis of α-ketols **5** and lactone **1**.

Conclusion

The biogenesis of solerone **1** and related compounds was successfully rationalized by biomimetic model reactions. As key step we established the pyruvate decarboxylase catalyzed acyloin condensation of pyruvic acid with ethyl 4-oxobutanoate **4** or ethyl 2-oxoglutarate **3** with acetaldehyde. The importance of the ethyl ester function in **3** and **4** serving as substrates for the enzymatic formation of α-hydroxy ketones **5** and **6** was demonstrated. The identification of six yet unknown sherry compounds including acyloins **5** and **6**, which have been synthesized for the first time, confirmed the relevance of the biosynthetic pathway. Application of MDGC-MS allowed the enantiodifferentiation of α-ketols and related lactones in complex sherry samples and disclosed details of their biogenetic relationship.

Literature cited

1. Schreier, P. *CRC Crit. Rev. Food Sci. Nutr.* **1979**, *12*, 59-71.
2. Augustyn, O. P.; Van Wyk, C. J.; Freeman, B. M.; Muller, C. J.; Kepner, R. E.; Webb, A. D. *J. Agric. Food Chem.* **1971**, *19*, 1128-1130.
3. Näf, R.; Jaquier, A.; Boschung, A. F.; Lindström, M. *Flav. Frag. J.* **1995**, *10*, 243-247.
4. Martin, B.; Etiévant, P.; Le Quéré, J.-L. *J. Agr. Food Chem.* **1991**, *39*, 1501-1503.
5. Fagan, G. L.; Kepner, R. E.; Webb, A. D. *Am. J. Enol. Vitic.* **1981**, *32*, 163-167.
6. Wurz, R. E. M.; Kepner, R. E.; Webb, A. D. *Am. J. Enol. Vitic.* **1988**, *39*, 234-238.
7. Berti, G.; Caroti, P.; Catelani, G.; Monti, L. *Carbohydr. Res.* **1983**, *124*, 35-42.
8. Georgiadis, M. P.; Haroutounian, S. A.; Apostolopoulos, C. A. *Synthesis* **1991**, *5*, 379-381.
9. Häring, D.; Schreier, P.; Herderich, M. *J. Agric. Food. Chem.* **1997**, *45*, 369-372.
10. Häring, D.; König, T.; Withopf, B.; Herderich, M.; Schreier, P. *J. High Resol. Chromatog.* **1997**, 351-354.
11. Lobell, M.; Crout, D. H. G. *J. Chem. Soc. Perkin Trans. 1* **1996**, 1577-1581.

12. Kren, V.; Crout, D. H. G.; Dalton, H.; Hutchinson, D. W.; König, W.; Turner, M. M.; Dean, G.; Thomson, N. *J. Chem. Soc. Chem. Commun.* **1993**, 341-343.
13. Suomalainen, H.; Konttinen, K.; Oura, E. *Arch. Mikrobiol.* **1969**, *64*, 251-261.
14. Muller, C. J.; Kepner, R. E.; Webb, A. D. *J. Agric. Food Chem.* **1972**, *20*, 193-195.
15. Muller, C. J.; Kepner, R. E.; Webb, A. D. *Am. J. Enol. Vitic.* **1973**, *24*, 5-9.
16. Heresztyn, T. *J. Agric. Food Chem.* **1984**, *32*, 916-918.
17. Herraiz, T.; Ough, C. S. *J. Agric. Food Chem.* **1992**, *40*, 1015-1021.
18. Meister, A. *J. Biol. Chem.* **1952**, *197*, 309-317.
19. Freeman, B. M.; Muller, C. J.; Kepner, R. E.; Webb, A. D. *Am. J. Enol. Vitic.* **1977**, *28*, 119-122.
20. Hollenbeak, K. H.; Kuehne, M. E. *Tetrahedron* **1974**, *30*, 2307-2316.
21. Bergman, J.; Lidgren, G. *Tetrahedron Lett.* **1989**, *30*, 4597-4600.
22. Mosandl, A.; Fischer, K.; Hener, U.; Kreis, P.; Rettinger, K.; Schubert, V.; Schmarr, H.-G. *J. Agric. Food Chem.* **1991**, *39*, 1131-1134.
23. Werkhoff, P.; Brennecke, S.; Bretschneider, W.; Güntert, M.; Hopp, R.; Surburg, H. *Z. Lebensm. Unters. Forsch.* **1993**, *196*, 307-328.
24. Bicchi, C.; Manzin, V.; D'Amato, A.; Rubiolo, P. *Flavour Fragrance J.* **1995**, *10*, 127-137.
25. Kaiser, R.E.; Rieder, R.I. *HRC & CC* **1987**, *10*, 240-243.
26. Bretschneider, W.; Werkhoff, P. *HRC & CC* **1988**, *11*, 543-546.
27. Guichard, E.; Etiévant, P.; Henry, R. Mosandl, A. *Z. Lebensm. Unters. Forsch.* **1992**, *195*, 540-544.
28. Hollnagel, A.; Menzel, E.-M.; Mosandl, A. *Z. Lebensm. Unters. Forsch.* **1991**, *193*, 234-236.

Chapter 10

Phenolic Composition as Related to Red Wine Flavor

Véronique Cheynier[1], Hélène Fulcrand[1], Franck Brossaud[1,2], Christian Asselin[2], and Michel Moutounet[1]

[1]Institut Supérieur de la Vigne et du Vin, INRA-IPV, Unité de Recherche Polymères et Techniques Physico-Chimiques, 2 place Viala, 34060 Montpellier cedex, France
[2]Unité de Recherches sur la Vigne et le Vin, 42, rue G. Morel, 49071 Beaucouzé cedex, France

Red wine quality and in particular color and flavor are largely related to phenolics, including both grape constituents and products formed during wine-making. The phenolic composition of Cabernet franc grapes harvested from various areas in the Loire Valley (France) and of the corresponding wines was determined and related to wine sensory properties. Grapes yielding intense and balanced wines were characterized by high anthocyanin to tannin ratios. Wine astringency increased with the level of procyanidins, and especially that of galloylated procyanidins but appeared less related to the prodelphinidin content, suggesting that these two groups of tannins, respectively abundant in seeds and specific of skins, have distinct organoleptic properties. It also depended on other constituents, including tannin polymers and tannin-anthocyanin adducts. Tannin quality seemed to depend on the relative amounts of both types of adducts, itself determined by the nature and proportions of competing precursors. A number of phenolic reaction products have been identified in wine. Along with polymerisation reactions which may participate in the flavor changes occurring as the wine ages, addition of various molecular species, including vinylphenol, pyruvic acid and acetaldehyde, to anthocyanins was demonstrated. These reactions result in color changes from purple to tawny and increased color stability. They may also contribute to lowering the level of volatiles and associated off-flavors in red wines.

Quality of red wines depends to a large extent on their phenolic composition. In particular, sensory analyses of wines obtained, over a fifteen-year period, from *Vitis vinifera* var. Cabernet franc grapes grown in different Loire Valley locations, pointed out to terroir-related characteristics presumably attributable to phenolic compounds (*1-3*). In fact, multiple factor analysis of the sensory data consistently showed the predominance of a first factorial axis associated with intensity variables (color, taste, flavor), mellowness and balance. Among these properties, color is obviously related to phenolics, as the red wine pigments consist of anthocyanins and their derivatives. Besides, taste attributes such as intensity, balance and mellowness are often considered related to wine phenolic composition. In particular, wine quality has been claimed to depend on anthocyanin to tannin ratio. Also, skin tannins are traditionally regarded by enologists as nicer or softer than seed tannins, although their structural differences were demonstrated only recently (*4*).

The organoleptic properties of tannins largely depend on their structures. In particular, low molecular weight flavanols, as well as gallic acid, are both bitter and astringent (*5*) and are likely to alter quality. Procyanidins become gradually less bitter and more astringent as the molecular weight increases up to about 10 units (*6*). Beyond this limit, they are believed to be insoluble and thus no longer astringent. Different perceptions of tannins in apple ciders, described as 'hard' (both bitter and astringent) or 'soft' (astringent but not bitter), have thus been interpreted in terms of balance of oligomeric to polymeric procyanidins (*6*). In the case of grape seed tannins, balance of bitterness and astringency appears concentration dependant too, bitterness being masked by greater astringency as the tannin content increases (*7,8*). This may be due to the particular structure of grape seed tannins, including galloylated procyanidins. The influence of galloylation and of B-ring trihydroxylation is not known. However, galloylation was shown to increase tannin interactions with various proteins, suggesting that it may also enhance astringency (*9,10*). Larger molecular weight tannins interacted more readily with proteins, thus protecting oligomers (*11*). Note that no size limit was observed in this experiment, although the larger molecular weight tannins tested contained 16 units in average. However, the extent of tannin-protein interactions may not reflect astringent perception.

Otherwise, acetaldehyde-induced polymerisation of proanthocyanidins has been said to participate in deastringency mechanisms during persimmon ripening (*12*). Similarly, the formation of tannin-anthocyanin adducts is commonly proposed to explain loss of astringency during wine-aging. In contrast, enzymatic inhibition studies indicated that interaction of products arising from catechin oxidation with enzymes was similar to that of procyanidins and in some cases higher (*13*). However, the influence of flavanol reactions on taste as well as their occurrence in wine remain speculative.

The purpose of our work was to determine eventual relationships existing between phenolic composition of grapes and wine and wine quality within the Loire valley parcel network. Detailed studies of grape and wine phenolic composition and of the various reactions involving grape polyphenols during wine-making and aging were therefore necessary.

Experimental

Chemicals. All solvents and acids used were analytical grade, except the MeOH and MeCN which were of HPLC grade, purchased from PROLABO (Fontenay-sous-bois, France). Malvidin 3-glucoside chloride and quercetin -3-glucoside were purchased from Extrasynthese (Lyon, France) and (+)-catechin and (-)-epicatechin from Sigma Chemical Co (Saint Louis, Mo). All other phenolic compounds were extracted from grape skins or seeds and purified by semi-preparative HPLC procedures as described earlier. *(4,14,15)*.

Polyphenol Analyses. Skin and seed extracts were prepared as described elsewhere (4,15). Flavonol and anthocyanin composition of grape skin extracts and wines were determined by direct reversed-phase HPLC analysis with diode array detection. The chromatographic conditions were the same as described earlier *(16)* but the formic acid concentration in the elution solvents was raised to 5% to improve anthocyanin resolution. Quantitations were based on peak areas, using malvidin-3-glucoside (at 530nm) and quercetin-3-glucoside (at 360 nm) response factors, respectively, for anthocyanins and flavonols.

Polymeric fractions were obtained from wines, seed and skin extracts by fractionation on a Toyopearl HW-40 column as described by Souquet *et al. (4)*. Two aliquots of the fractions containing polymeric material were taken to dryness under vacuum. The first one was used to determine proanthocyanidin composition by thiolysis followed with HPLC analysis *(17)*. The other one was dissolved in MeOH acidified with 2% HCl and used to estimate the concentration of total polymeric polyphenols and polymeric pigments by measuring the absorbance, respectively at 280 nm and 530 nm. Absorbance data were converted to equivalent epicatechin and equivalent malvidin-3-glucoside, respectively, using the extinction coefficients determined for each compounds under similar conditions.
All analyses were performed in triplicate.

Grapes and wines. Wines were prepared at the INRA experimental station at Beaucouzé from *Vitis vinifera var.* Cabernet franc grapes from different sites in the Loire Valley region (France) in 1995. Fifteen wines were made from ten Saumur parcels (coded CHA, DAM, 2EL, 3EL, 4EL, FON, ING, PER, POY, VAU) and 5 Anjou parcels (1AL, 2AL, FAL, GRA, SCI) harvested at commercial maturity, starting each time from 80 kg grape samples. Two additional wines (coded ING2 and VAU2) were prepared from grapes harvested two weeks later from parcels ING and VAU known for their lack of earliness. Wine-making was carried out in stainless steel tanks, with 8 days pomace contact. Each tank was inoculated with 0.2 g/L yeast (*Saccharomyces cerevisiae*, INRA Narbonne-7013 strain) and fermented to dryness (less than 2g/L residual sugars) at 25°C. Wines were then maintained at 20°C until the end of malolactic fermentation. Three-hundred berries were randomly taken from Saumur samples, including VAU2 and ING2, and stored at -20 C for chemical analyses.

Sensory assessment. After six months, wines were tasted by a panel of 30 experienced judges who had been trained for tannin tasting using the terms : soft tannins, green tannins, drying tannins, hard tannins, tannin quality. Wine organoleptic profiles were analyzed using the following 20 descriptors allocated to vision (intensity, tint), odor (intensity, fruity, veggy, smoky and animal characters), flavor (intensity, persistence) taste (intensity, frankness, acidity, alcohol, mellowness, balance), and tannin characteristics listed above, which were noted from 1 to 5. The data thus obtained was averaged into a single value for each characteristic and each wine.

Statistical analysis. All data obtained (chemical data and sensory data) were treated by Multiple Factor Analysis (MFA) *(2)* using the ADDAD software (ADDAD 89, Escofier B. & Pagès J., France).

Phenolic Composition of Grapes and Wines.

Grape Polyphenol Composition. Grape polyphenols show a great diversity of structures and properties. They include flavonoids, based on a C6-C3-C6 skeleton, and non flavonoids, mostly represented by phenolic acids (*i.e.* benzoic acids and hydroxycinnamic acids) and stilbenes, Figure 1. Among flavonoids, anthocyanins and flavanols are particularly important to wine quality, as they are respectively red grape pigments and grape tannins. In contrast, phenolic acids are colorless and tasteless in buffered media such as wine *(18)*, with the exception of gallic acid, which is both bitter and astringent *(5)*. However, hydroxycinnamic acids, as well as flavanols, may proceed to brown and potentially astringent products via oxidative reactions *(19)*. Molecular size increase of the oxidation products will ultimately lead to precipitation and reduction of astringency *(20)*.

Each phenolic class comprises various structures, differing by the number and position of hydroxy groups, which can also be diversely substituted (*e.g.* glycosylated, acylated...). For example, anthocyanins are encountered in *Vitis vinifera* as the 3-glucosides of cyanidin (R=H, R'=OH), peonidin (R=H, R'=OCH₃), petunidin (R=OH, R'=OCH₃), delphinidin (R=R'=OH) and malvidin (R=R'=OCH₃) (cf Figure 1), along with the corresponding acetyl-, *p*-coumaroyl- and caffeoyl-glucosides. Malvidin-3-glucoside and its derivatives are always the predominant species but varietal differences in the anthocyanin amounts and relative proportions are observed *(21)*. Besides, grape flavonoid composition is also influenced by environmental factors such as temperature, sun exposure, and growing area (22-24).

Flavanols similarly exist as diversely hydroxylated and/or substituted monomeric species but also as oligomers and polymers, called condensed tannins or, because they release anthocyanidins when heated in acidic medium, proanthocyanidins.

Several proanthocyanidin classes can be distinguished on the basis of the hydroxylation pattern of the constitutive units. Among them, procyanidins, consisting of (epi)catechin units (3', 4' di OH), and prodelphinidins, deriving from (epi)gallocatechin (3',4',5' tri OH), Figure 2, have been reported in grapes.

Monomeric units may be linked, to form oligomers and polymers, by C4-C6 and/or C4-C8 bonds (B-type) or doubly linked, with an additional C7-O-C2 ether linkage (A-type). Besides, flavan-3-ol units may be encountered as 3-O-esters, in particular with gallic acid, or as glycosides *(25)*. Finally, the degree of polymerization (DP) may vary greatly as proanthocyanidins have been described up to 20,000 in molecular weight *(26)*.

Only B-type proanthocyanidins have been formally identified in grapes, with small amounts of dimers and trimers containing 4-6 linkages occuring along with the

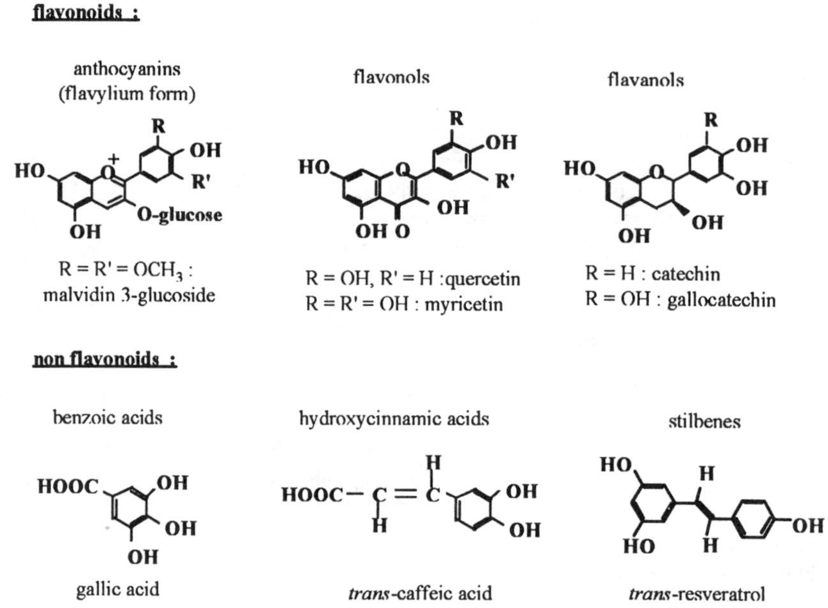

flavonoids :

anthocyanins
(flavylium form)

R = R' = OCH$_3$:
malvidin 3-glucoside

flavonols

R = OH, R' = H :quercetin
R = R' = OH : myricetin

flavanols

R = H : catechin
R = OH : gallocatechin

non flavonoids :

benzoic acids

gallic acid

hydroxycinnamic acids

trans-caffeic acid

stilbenes

trans-resveratrol

Fig. 1 : Structures of phenolic compounds

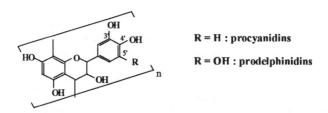

R = H : procyanidins

R = OH : prodelphinidins

Fig. 2. Structure of grape proanthocyanidins

most common 4-8 linked oligomers *(27)*. Seed tannins are partly galloylated procyanidins, based on (+)-catechin, (-)-epicatechin and (-)-epicatechin 3-O-gallate units (15), whereas skin tannins also contain prodelphinidins, detected as (-)-epigallocatechin, along with trace amounts of (+)-gallocatechin and (-)-epigallocatechin 3-O-gallate *(4)*. Grape seeds show larger amounts of tannins and larger proportions of galloylated units than grape skins whereas the average molecular weight is higher in skins than in seeds.

Wine Phenolic Composition. Wine phenolic composition depends on the grape from which the wine is made but also on the wine-making conditions which influence extraction of the various compounds from grape and their subsequent reactions.

Diffusion kinetics. The diffusion kinetics vary greatly among polyphenols, due to differences in localization within the grape berry and in solubility. Anthocyanins are readily extracted from skins during the first days of pomace contact and then gradually degraded *(28)* whereas flavanol concentration continues to increase for about two weeks. Among the latter, prodelphinidins diffuse faster than procyanidins and especially galloylated procyanidins, owing either to their larger accessibility (presence in skins) or higher hydrophilicity. Larger molecular weight tannins also diffuse later than smaller oligomers *(11)*.

Phenolic Reactions in Wine. Polyphenols are extremely unstable compounds. Their reactions start as soon as the grape is crushed or pressed and continue throughout wine-making and aging, leading to a great diversity of new products. These products show specific organoleptic properties, often different from those of their precursors. Therefore, better understanding of their structures and of the mechanisms responsible for their formation appears necessary to predict and control wine quality.

Thus, continuous color changes from purple to tawny and increasing color stability towards pH variations and sulfite bleaching observed during aging of red wines are due to conversion of grape anthocyanins to other pigments *(29)*. As well, astringency decrease results from reactions (degradation, polymerisation, addition) of tannins *(20)*.

Anthocyanin reactions are classically described as anthocyanin-tannin additions, which can either be direct, generating orange xanthylium salts, or involve acetaldehyde, leading to purple pigments. Other tannin reactions are of two major types : on one hand, acid-catalysed bond-making and bond-breaking processes characteristic of proanthocyanidin chemistry *(20)*, on the other hand, oxidation reactions leading to browning *(19,30,31)*.

Some of these reactions and especially those involving anthocyanins, flavanols, and acetaldehyde *(32-37)*, have been thoroughly studied in wine-like model systems *(32-43)*. Numerous products have thus been obtained and partly characterized. Besides, some of them have recently been detected in red wines *(38, 44)*. Two different groups of reactions were thus shown to occur in the course of wine making.

The first one is the classical acetaldehyde-induced tannin-anthocyanin addition described in the literature. Detection of a catechin-ethanol adduct by LC-

MS *(38)* demonstrated that the reaction starts with protonation of acetaldehyde in acidic medium, followed by nucleophilic addition of the flavanol (C6 or C8 of the A-ring) on the resulting carbocation, Figure. 3, as postulated by Timberlake and Bridle *(32)*. The ethanol adduct then looses a water molecule to give a new carbocation which is in turn attacked by the anthocyanin to give the flavanol-ethyl-anthocyanin adduct.

The anthocyanin can be replaced in this process by another flavanol molecule so that formation of ethyl-linked flavanol polymers, Figure 4, competes with that of flavanol-ethyl-anthocyanin condensation products, Figure 4 (2). In flavanols, the C6 and C8 positions seem equally reactive. Successive condensations thus lead to numerous oligomers and polymers, in which constitutive units are linked by ethyl bridges. However, it seems that the reaction stops when both ends are occupied by anthocyanin units *(37)*. A great diversity of products can thus be generated during wine aging, their respective levels depending on the nature and relative amounts of flavanols and anthocyanins present.

Moreover, it was shown that acetaldehyde can also be replaced by other aldehydes in this reaction. In particular, reaction of flavanols with glyoxylic acid resulting from oxidation of tartaric acid yielded other types of polymers linked through carboxymethine bridges *(37,43)*, Figure 4 (3).

The second mechanism demonstrated consists in a cycloaddition between anthocyanins and various wine components possessing a polarisable double bond *(45)*. These include in particular 4-vinylphenol *(41)* and several yeast metabolites such as acetaldehyde and pyruvic acid, Figure 5. Since some of these products are volatile, their conversion to non volatile anthocyanin-adducts is likely to modify wine odor ; in particular, reaction of vinylphenol with anthocyanin may contribute to lowering its concentration and associated off-flavors in red wines *(41)*. Besides, the products formed by these reactions show a red-orange color and are exceptionally stable towards pH variations and sulfite action *(45,46)*. Although they represent only a small proportion of young wine pigments, their concentration remains constant as the wine ages so that they gradually become predominant amongst monomeric pigments. Moreover, the similarities of the color properties of these new anthocyanin adducts with those of old wine pigments suggest that they are based on analogous structures and result from similar reactions. In fact, mass and UV-visible spectra of another product formed from procyanidin dimer B2 and malvidin-3-glucoside in the presence of acetaldehyde *(47)* indicated that proanthocyanidins may also participate in these reactions, which thus presumably lead to a whole range of polymeric pigments.

Phenolic Composition as Related to Wine Flavor : the Loire Valley Parcel Network.

Wines made from *Vitis vinifera* var. Cabernet franc grapes harvested from different Loire Valley locations were submitted to sensory evaluation. The grapes and wines were also analysed for their phenolic composition. The three sets of data thus generated were then treated separately by mutiple factor analysis (MFA) in order to compare the wine configurations thus obtained and establish eventual

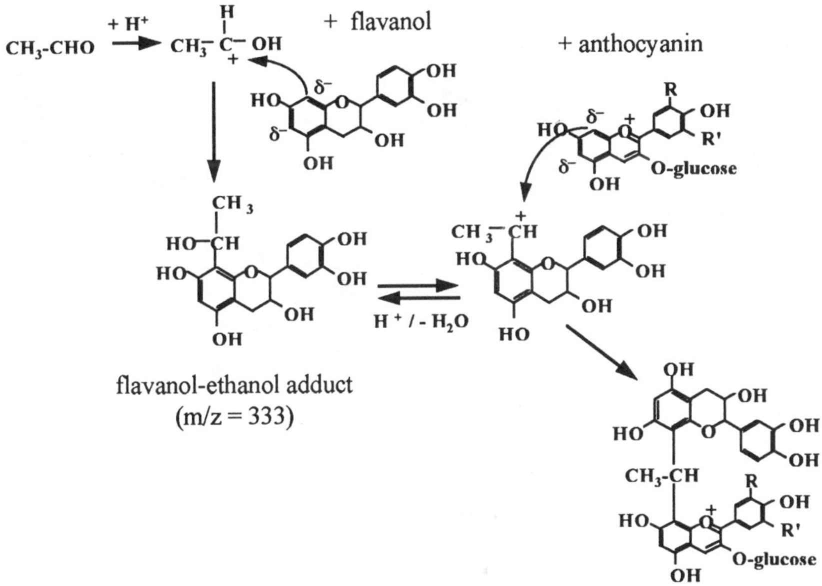

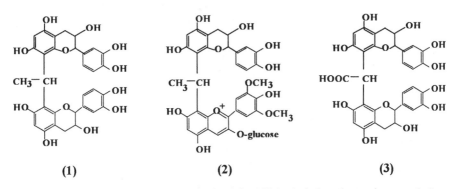

Figure 3 : acetaldehyde-induced flavanol-anthocyanin condensation

Figure 4 : examples of products obtained by aldehyde-induced reactions : ethyl-linked catechin dimer (1), ethyl-linked catechin-malvidin-3-glucoside adduct (2), ethanoic acid-linked catechin-dimer (3)

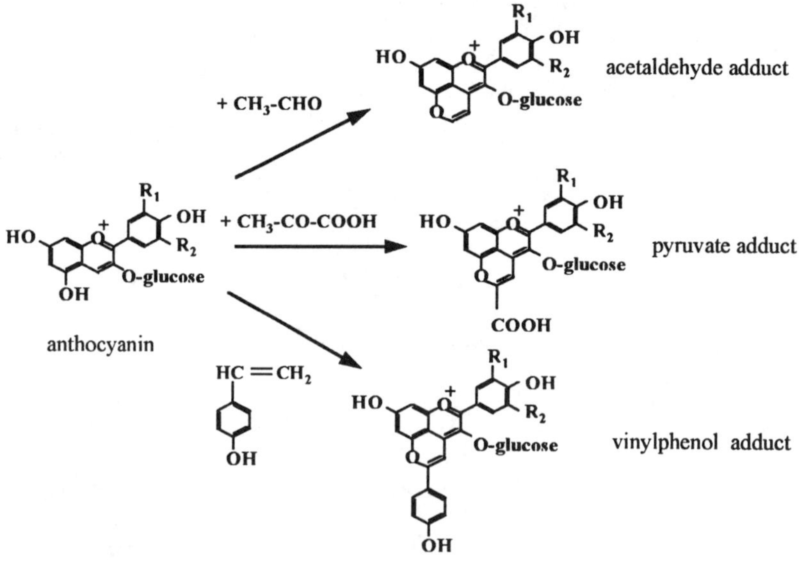

132

Figure 5. some addition reactions of anthocyanins in wine

relationships between sensory attributes of the wines and chemical composition of raw material or processed wines.

Sensory Analysis of the Saumur Wines. Multiple factor analysis of the data established by sensory evaluation of the 1995 Saumur wines, Figure 6, showed that the first axis, accounting for 72% of the total variance, contrasted 'green' tannins against color, flavor, taste and aroma intensities, flavor persistence, tannin quality, 'soft' tannins, mellowness and balance which were all highly correlated together. Projection of the samples along this axis allowed us to distinguish wines made from ING, ING2 and VAU parcels, from other wines showing larger intensity and balance, in agreement with earlier observations *(1-3)*.

Chemical Analysis of the Saumur Grapes and Wines. MFA of the grape phenolic composition, Figure 7, gave approximately the same sample distribution as the sensory data along the first axis (45 % of the total variance) which contrasted anthocyanins against seed tannins and procyanidin gallates. The second axis, representing 25 % of the total variance was defined by prodelphinidins and skin tannins but it could not be related to sensory differences.

Similar sample distribution was obtained by MFA analysis of the wine analytical data, with the exception of VAU2 which was different from VAU by sensory analysis of the wines and very close to it with regards to phenolic composition. The first axis contrasted the concentrations of anthocyanins, including polymeric pigments, against those of proanthocyanidins (tannins), gallates, and total polymers, which were highly correlated together.

Thus, higher quality Cabernet franc wines, characterized by high intensity and mellowness, were obtained from grapes showing high ratios of anthocyanins to seed tannins and galloylated procyanidins. In contrast, 'green' tannin perception was associated with low ratios of anthocyanin to tannins in grapes. The same composition differences were also observed in the corresponding wines. Besides, higher quality wines contained larger amounts of polymeric pigments and lower amounts of total polymers. This suggests that the reaction pathways leading to both types of polymers are in competition so that their relative importance is determined by the anthocyanin to tannin ratio.

Sensory Analysis of the Saumur and Anjou Wines as Related to Phenolic Composition. Sensory analysis of the Saumur and Anjou wines, Figure 8, allowed to distinguish a third group of wines (namely SCI and GRA), characterized by strong astringency, in addition to the higher quality and poorer quality wines mentioned above.

Each of the three groups was associated with specific tannin descriptors, namely 'green' tannins for the poorer quality wines, 'soft' tannins for the higher quality wines, 'hard' and 'drying' tannins for the astringent wines. Comparison of the wine distribution obtained respectively from the sensory data, Figure 8, and compositional data, Figure 9, indicated that astringency was associated to high levels of tannins, including in particular gallates but also prodelphinidins, of total polymers, and to a lesser extent of polymeric pigments, whereas 'soft' tannins were associated with high levels of anthocyanins. Wines described as 'green' were not particularly rich in tannins, but they were characterized by a lack of anthocyanins

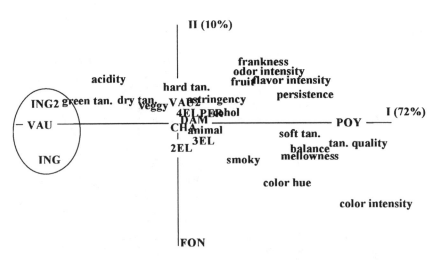

Figure 6 : MFA of the Saumur wine sensory data : projection of the sensory variables (lower case) and of the wines (capitals) on the first and second factors

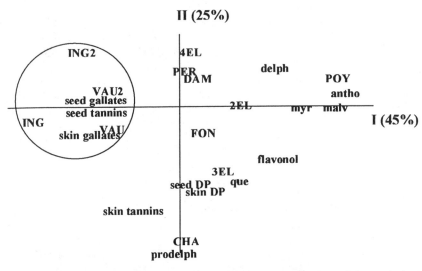

Figure 7 : MFA of the Saumur grape analytical data : projection of the analytical variables (lower case) and of the wines (capitals) on the first and second factors.

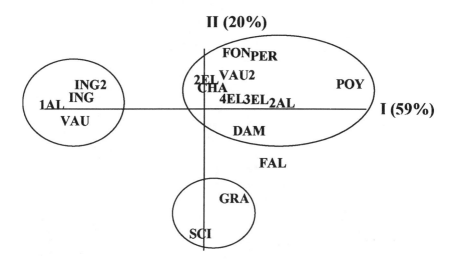

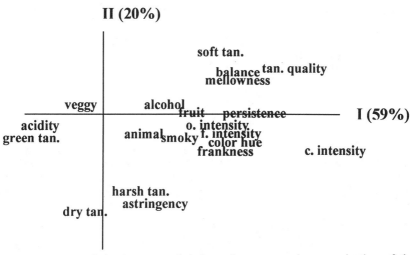

Figure 8 : MFA of the Saumur and Anjou wine sensory data : projection of the wines (upper part) and of the variables (lower part) on the first and second factors.

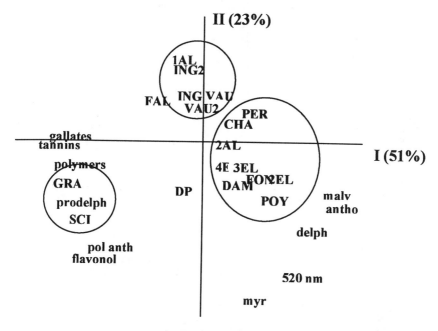

Figure 9 : MFA of the Saumur and Anjou wine analytical data : projection of the wines (capitals) and of the variables (lower case) on the first and second factors.

and especially of polymeric pigments compared to higher quality wines. The 'green' tannin character was also associated with sourness by the tasting panel, and attributed to a lack of maturity of the grapes. However, this was not related to wine acidity, as pH and titratable acidity values measured for the 'sour' wines were in the same range as those of high quality wines. Astringent wines were in fact the most acidic wines, with lower pH values and higher total acidity values than all others. This suggests that high astringency masked acidity perception. Inversely, high acidity has also been shown to enhance the intensity of astringency (48). The differences in tannin quality did not seem to be related either to the alcohol content, although raising ethanol concentration was earlier shown to increase bitterness and decrease astringency (48). Note however that the higher alcohol content measured in VAU2 (13% ethanol) compared to VAU (10.2% ethanol) may explain why the former appeared much better than the latter in the sensory analysis, although they had similar phenolic composition. Finally, the possibility that the panellists' perception of tannins has been influenced by wine color cannot be ruled out.

Discussion

Relationships between tannin composition and taste. Our results indicate that astringency, defined as an extreme drying or puckering sensation within the mouth resulting from interactions between tannins and mouth proteins (49,50) is essentially due to procyanidins, and especially galloylated procyanidins, although the most astringent wines also contained high levels of prodelphinidins and other polymeric material. This is in agreement with model solutions studies demonstrating that tannin interactions with various proteins increase with the extent of galloylation both in the case of hydrolysable tannins (26) and in that of procyanidins (9,10). Prodelphinidins, unlike procyanidin gallates, did not seem to contribute to the 'green' tannin character, suggesting that B-ring trihydroxylation confers particular taste properties. Although the structure-taste relationships explaining such perception differences remain unknown, they can be related to the commonly acknowledged higher quality of skin tannins (containing prodelphinidins and low proportion of galloylated units) as opposed to seed tannins (consisting of procyanidins, with 30% galloylated units). Note that the proportions of galloylated units and epigallocatechin units constituting wine proanthocyanidins varied from 4 to 9 % and from 7 to 18%, respectively, in the Cabernet franc wines studied.

    Another interesting point is that the taste attribute described as 'soft' tannins, highly correlated with tannin quality (0.86), mellowness (0.91) and balance (0.83), corresponded to low tannin and high anthocyanin levels. Since anthocyanins are tasteless (50), this may mean that small amounts of tannins contribute positively to quality.

    According to Noble and coworkers, seed tannins are perceived as rather bitter at low concentrations, but astringency takes over as the concentration increases (7,8). Otherwise, in apple ciders, 'hard' tannin perception has been ascribed to bitterness (6). Nevertheless, in the Cabernet franc wines studied, low proanthocyanidin content appeared associated with 'soft' tannins. Such discrepancy may be due to the particular composition of wine tannins, including prodelphinidins and various tannin-like structures (e.g. oxidation products, ethyl-linked tannin polymers, anthocyanin-tannin adducts...), in addition to seed tannins.

Influence of anthocyanins on tannin perception. Another hypothesis is that interactions and/or reactions with anthocyanins alter tannin perception in wines showing high anthocyanin to tannin ratios. For example, increased tannin solubility resulting from complexation with anthocyanins *(50)* may prevent them from interacting with salivary and buccal epithelium proteins. Besides, tannin reactions yield different products in the presence of anthocyanins. In fact, formation of tannin-anthocyanin adducts compete with tannin polymerisation reactions, as both types of species show nucleophilic properties and may therefore participate in addition reactions leading to tannin derived products. Consequently, the nature and relative amounts of such products in wine depend on their anthocyanin to tannin ratio. Our experiments suggest that formation of anthocyanin-tannin adducts is the major mechanism involved in the conversion of astringent proanthocyanidins (hard tannins) to soft tannins during wine maturation, in agreement with earlier studies *(20,29)*.

However, flavanol reactions may also contribute to taste changes. In particular, acetaldehyde-induced polymerisation may participate in the deastringency process, as postulated earlier in the case of persimmon ripening. In contrast, colorless catechin oxidation products were shown to interact with enzymatic proteins as did procyanidin dimers *(13)*, and should therefore be similarly astringent. Further reaction of the primary oxidation products increased the inhibitory effect of the solution, suggesting that the new pigment species formed may be more astringent than their precursors. Nevertheless, the increase in astringency resulting from oxidative polymerization was earlier said to be accompanied with a softening effect, due to loss of bitterness *(48)*. Besides, catechin quinones generated by oxidation are likely to proceed to different - and possibly 'softer'- products in the presence of anthocyanins, as demonstrated earlier in the case of caffeoyltartaric acid quinones *(51)*.

Conclusion

The data presented point out the complexity of wine phenolic composition and the difficulties encountered to relate it to sensory properties.

Grape tannins consist essentially of proanthocyanidins, which are partly galloylated procyanidins in seeds, procyanidins and prodelphinidins in skins. Wine tannin composition is influenced by the fermentation conditions (*e.g.* skin contact duration, temperature) as rather polar tannins from skins (prodelphinidins) are more readily extracted than less polar –and presumably also less accessible– seed components (galloylated procyanidins). In addition to grape proanthocyanidins, wine tannins include non-proanthocyanidin tannin-like structures, formed in increasing amounts as the wine ages. Trihydroxylated compounds (*i.e.* prodelphinidins and gallates) and larger molecular weight proanthocyanidins seem to proceed to such derivatives faster than oligomeric procyanidins. However, the various types of proanthocyanidins may undergo different reactions, and certainly yield different products, showing distinct organoleptic properties.

Comparison of the chemical and sensory data obtained when analysing 17 Cabernet franc wines from the Loire Valley suggested that procyanidin gallates and prodelphinidins show different tannin characters although the observed differences may also be related to tannin content and/or interferences with other wine constituents. Low levels of procyanidins associated with high concentrations of anthocyanins seemed to contribute to mouthfullness (attributed to 'soft' tannins), which was also possibly enhanced by alcohol. As the amount of tannins increased and the ratio of anthocyanin to tannin decreased, tannin perception shifted from soft to green and then from green to hard and drying. However, astringent sensation may also have been enhanced by acidity. Tannin 'softness' as opposed to 'hardness' appeared also related to the respective levels of anthocyanin-tannin adducts and of tannin-based polymers. Since competition between pathways leading to both types of products is governed by the initial anthocyanin to tannin ratio, this may explain why high levels of anthocyanins are necessary to obtain not only visual intensity but also balance and mellowness. Finally, 'green' tannins characteristic of poor quality Cabernet franc wines, were associated to lack of anthocyanins and polymeric pigments, suggesting that anthocyanins participate in wine flavor, although the panelists may have been influenced by the poorer color of the wines.

Much further work remains to be done to determine the reaction mechanisms involving phenolic compounds, their importance relative to each other in wine, and the nature and properties of the resulting products. Likewise, studies are needed to predict the actual taste of the various proanthoanthocyanins and of their numerous derivatives but also eventual synergistic or antagonist effects. In particular, the role of anthocyanins has to be investigated.

Acknowledgments.

Financial support and material support from The Conseil Interprofessionnel des Vins d'Anjou et de Saumur (Angers, France) and from the Syndicats des vins rouges d'Anjou et de Saumur is greatfully acknowledged.

Literature Cited.

1.   Morlat, R.; Asselin, C.; Pagès, J.; Leon, H.; Robichet, J.; Remoue, M.; Salette, J.; Sigogne, M. *Connaiss. Vigne Vin* **1983**, *17*, 219-246.
2.   Pagès, J.; Asselin, C.; Morlat, R.; Robichet, J. *Sci. Aliments* **1987,** *7*, 549-571.
3.   Asselin, C.; Pages, J.; Morlat, R. *J. Int. Sci. Vigne Vin* **1992,** *26*, 129-154.
4.   Souquet, J.M.; Cheynier, V.; Brossaud, F.; Moutounet, M. *Phytochemistry*, **1996,** *43*, 509-512.
5.   Robichaud, J. L.; Noble, A. C. *J. Sci. Food Agric.* **1990**, *53*, 343-353.
6.   Lea, A.G.H. In *Bitterness in foods and beverages,* Rouseff, R.L. Ed. *Developments in Food Science, Elsevier,* **1990**, *25*, 123-143.
7.   Singleton, V.L.; Noble, A.C. In *Phenolic Sulfur and Nitrogen Compounds in Food Flavors;* Charalambous, G.; Katz, I., Eds.; *A.C.S. Symp. Series,* **1976**, *26*, 47-70.
8.   Arnold, G.M.; Noble, A.C. *Am. J. Enol. Vitic.* **1978**, *29*, 150-152.

9. Cheynier, V.; Rigaud, J.; Ricardo da Silva, J.M. In *Plant Polyphenols.* Hemingway, R.W.; Laks, P.E., Eds.; Plenum Press, New York, **1992**, 281-294.

10. Ricardo da Silva, J.M.; Cheynier, V.; Souquet, J.M.; Moutounet, M.; Cabanis, J.C.; Bourzeix, M.; *J. Sci. Food Agric.* **1991**, *57*, 111-125.

11. Cheynier, V.; Prieur, C.; Guyot,S.; Rigaud, J.; Moutounet, M. *in* "Wine composition and health benefits", T. Watkins, ed..*A.C.S. Symp. Series,* **1997**, *661*, 81-97.

12. Tanaka, T.; Takahashi, R.; Kouno, I.; Nonaka, G.I. *J. Chem Soc. Perkin Trans I,* **1994**, 3013-3022.

13. Guyot, S.; Pellerin, P.; Brillouet, J.-M.; Cheynier, V. *Biosci. Biotechnol. Biochem.,* **1996**, *60*, 1131-1135.

14. Sarni, P.; Fulcrand, H.; Souillol, V.; Souquet, J.M.; Cheynier, V. *J. Sci. Food Agric.,* **1995**, *9*, 385-391.

15. Prieur, C.; Rigaud, J.; Cheynier, V.; Moutounet, M. *Phytochemistry* **1994**, *36*, 781-784.

16. Cameira dos Santos, P.J.; Brillouet, J.M.; Cheynier, V.; Moutounet M. *J. Sci. Food Agric.* **1996**, *70*, 204-208.

17. Rigaud, J.; Perez-Ilzarbe, J.; Ricardo da Silva, J.M.; Cheynier, V. *J. Chromatogr.* **1991**, *540*, 401-405.

18. Vérette, E.; Noble, A.C.; Somers, T.C. *J. Sci. Food Agric.* **1988**, *45*, 267-272.

19. Cheynier,V.; Fulcrand, H.; Guyot, S.; Oszmianski, J.; Moutounet, M. in *Enzymatic Browning and its Prevention;* Lee C.Y.; Whitaker, J.R., Eds.; *ACS Symp. Series* **1995**, *600*, 130-143.

20. Haslam, E. *Phytochemistry* **1980**, *19*, 2577-2592.

21. Mazza, G.; Miniati, E; In : *Anthocyanins in fruits, vegetables and grains.* C.R.C. Press. Boca Raton, Ann Arbor, London and Tokyo, 1993, 149-199.

22. Price, S.; Breen, P.J.; Vallado, M.; Watson, B.T. *Am. J. Enol. Vitic.,* **1989**, *46*, 187-194.

23. Larice, J.L., Archier, P.; Rocheville-Divorne, C.; Coen, S.; Roggero, J.P. *Rev. F. Oenol.* **1989**, *121,* 7-12.

24. Rigaud, J.; Cheynier, V.; Asselin, C.; Brossaud, F.; Moutounet, M. In *Oenologie 95,* ed. Lavoisier, Londres, Paris, New York, 1996, 137-140.

25. Porter, L.J. In *The Flavonoids. Advances in Research since 1980* Harborne, J.B., Ed.; Chapman and Hall, London, 1988, 21-62.

26. Haslam, E.; Lilley, T.H. *Crit. Rev. Food Sci. Nutr.* **1988**, *27*, 1-40.

27. Ricardo da Silva J.; Rigaud, J.; Cheynier, V.; Cheminat, A.; Moutounet, M. *Phytochemistry,* **1991**, *30*, 1259-1264.

28. Cheynier, V.; Hidalgo-Arellano, I.; Souquet, J.M.; Moutounet, M. *Am. J. Enol. Vitic.* **1997**, *48*, 225-228

29. Somers, T.C. *Phytochemistry,* **1971**, *10*, 2175-2186.

30. Guyot, S.; Cheynier, V.; Souquet, J.M.; Moutounet, M. *J. Agric. Food Chem.,* **1995**, *43*, 2458-2462.

31. Oszmianski, J.; Cheynier, V.; Moutounet, M. *J. Agric. Food Chem.,* **1996**, *44*, 1972-1975.

32. Timberlake, C.F.; Bridle P. *Am. J. Enol. Vitic.* **1976**, *27*, 97-105.

33. Baranowski, E.S.; Nagel, C.W. *J. Food Sci.* **1983**, *48*, 419-421.

34. Bakker, J.; Picinelli, A.; Bridle, P. *Vitis,* **1993**, *32*, 111-118.

35. Dallas, C.; Ricardo da Silva, J.M.; Laureano, O. *J. Sci. Food Agric.* **1995**, *70*, 493-500

36. Rivas-Gonzalo, J.C.; Bravo-Haro, S.; Santos-Buelga, C. *J. Agric. Food Chem.* **1995**, *43*, 1444-1449.

37 Fulcrand, H.; Es-Safi, N.; Doco, T.; Cheynier, V.; Moutounet, M. *Polyphenol communications 96,* **1996**, 202-203.

38. Fulcrand, H., Doco, T., Es Safi, N.; Cheynier, V. *J. Chromatogr.,* **1996**, 752, 85-91.

39. Liao, H.; Cai, Y.; Haslam, E. *J. Sci. Food Agric.* **1992**, *59*, 299-305.

40. Cameira dos Santos, P.J.; Brillouet, J.M.; Cheynier, V.; Moutounet, M. *J. Sci. Food Agric.* **1996**, *70*, 204-208.

41. Fulcrand, H.; Cameira dos Santos, P.J.; Sarni-Manchado, P.; Cheynier, V.; Favre-Bonvin, J. *J. Chem. Soc. Perkin Trans I,* **1996**, 735-739.

42. Fulcrand, H.; Cheynier, V.; Oszmianski, J; Moutounet, M. *Phytochemistry,* **1997**, *46*, 223-227.

43. Guyot, S.; Vercauteren, J.; Cheynier, V. *Phytochemistry* **1996** *42*, 1279-1288.

44. Bakker, J.; Timberlake, C.F.*J. Agric. Food Chem,* **1997**, *45*, 35-43.

45. Cheynier, V.; Fulcrand, H.; Sarni-Manchado, P.; Cameira dos Santos, P.J.; Moutounet, M. FP 96-00790/1995, extended to PCT countries.

46. Sarni-Manchado, P.; Fulcrand, H.; Souquet, J.M.; Cheynier, V.; Moutounet, M. *J. Food Sci.,* **1996**, *61*, 938-941.

47. Francia-Aricha, E.M.; Guerra, M.T.; Rivas-Gonzalo, J.C.; Santos-Buelga, C. *J. Agric. Food Chem.,* **1997**, *45*, 2262-2266.

48. Noble, A.C. In *Bitterness in foods and beverages,* Rouseff, R.L. Ed. *Developments in Food Science, Elsevier,* **1990**, *25*, 145-158.

49. Joslyn, M.A.; Goldstein, J.L. *Adv. Food Res.* **1964**, *13*, 179-219.

50. Singleton, V.L.; Trousdale, E.K. *Am. J. Enol. Vitic.* **1992**, *43*, 63-70.

51. Sarni-Manchado, P.; Cheynier, V.; Moutounet, M. *Phytochemistry,* **1997**, *45*, 1365-1369.

Chapter 11

Effects of Small-Scale Fining on the Phenolic Composition and Antioxidant Activity of Merlot Wine

Jennifer L. Donovan, Julie C. McCauley, Nuria Tobella Nieto, and Andrew L. Waterhouse[1]

Department of Viticulture and Enology, University of California, One Shields Avenue, Davis, CA 95616-8749

Fining is carried out on wine by adding one (usually insoluble) substance to remove one or more undesirable components. The levels of phenolic compounds, such as tannins, in wine are often reduced by the addition of proteinaceous or synthetic fining agents. Here, changes in the levels of total phenol and monomeric phenolic compounds were quantified after the addition of common fining agents. The proteins had modest effects on most monomeric compounds, but PVPP, a synthetic protein-like polymer, greatly reduced some compounds, especially quercetin and the resveratrols. Also, the proteins had little effect on the level of total phenol, while carbon and PVPP caused significant reductions. Unexpectedly, bentonite, a clay fining agent typically used to remove proteins, reduced anthocyanin levels, as well as the level of total phenol. When diluted to the same concentration (5 µM) of total phenol, the PVPP-treated wine was markedly more potent antioxidant for LDL. This change in specific antioxidant activity may be caused by differential changes in the tannin composition.

Fining is carried out to reduce the levels of certain wine components such as undesirable flavors and colors and to improve the clarity or long-term stability of wine. Other factors to consider are the effects of fining on browning or oxidation, the "unmasking" of undesirable flavors, undesirable reduction of color or flavor, the precipitation of the remaining fining agent upon aging (1), and changes in the filterability of the wine (2).

One of the major targets of fining agents are the phenolic compounds which are responsible for color, astringency, and bitterness, and may contribute to the body of the wine (3). In addition to these wine characteristics, the phenolic components are thought to be responsible for the reduced incidence of death due to coronary heart disease seen in populations who consume moderate amounts of wine (4). These compounds are thought

[1]Corresponding author.

to function as antioxidants for low density lipoproteins (LDL) (5). LDL oxidation has been shown to be a critical and necessary step in the development of atherosclerosis, and prevention of this step is thought to slow the progression of the disease (6,7). Phenolic compounds may also affect heart disease by effects on platelet aggregation (8).

Proteinaceous fining agents are often used to "soften" or reduce the astringency of the wine. One mechanism of interaction between the proteinaceous fining agent and phenolic compounds is by hydrogen bonding between the phenolic hydroxyl and the carbonyl oxygen of the peptide bond (9). The capacity of a protein fining agent is partially a function of the number of potential hydrogen bonding sites per unit weight and the accessibility or exposure to the sites. Protein binding increases as the size of the phenolic compound increases, so binding increases as the number of flavan-3-ol units increases (10-14). Protein fining did not affect dimeric and trimeric procyanidin levels in red wines, although total phenol levels were reduced (15). Interestingly, monomeric and dimeric flavonoids have been shown to interfere with tannin precipitation by protein (1).

Studies comparing different proteins have shown that egg whites generally have a larger effect on phenolic compounds than gelatin, reducing the content of leuco-anthocyanins and tannins, and decreasing the color in red wines (16,17). Hagerman and Butler (1981) showed that the affinity for tannins is an inverse function of the size of the protein, and peptides with less than six residues interact very weakly with tannin (9). In the larger-sized proteins, there are also considerable differences in the interactions of phenolics with different molecular weight fractions of gelatin. Yokotsuka and Singleton (1987) (1) showed that the smallest protein tested (2000 MW) was the most efficient in precipitating tannins, probably because of a higher hydrophobic character (proline content) and more similar sizes between the phenol and the peptide. The amino acid composition of proteins has an effect, and notably increased proline and hydroxyproline increases the affinity towards polymeric phenols (9,15,18).

PVPP, polyvinyl polypyrrolidone, is a synthetic fining material which tends to bind the monomeric and small polyphenols due to a conformational preference that permits hydrogen bonding of the carbonyl groups on PVPP and the phenolic hydrogens (12,19-21). Fining with PVPP preferentially removes smaller components that may be associated with bitterness, browning, and color (10,17,21-23). Ough (1960) also found that PVPP removed more tannin and color than gelatin in red wines (24).

Agar, the most commonly-used polysaccharide for fining, is a long-chained polymer of beta-1,4-D-manuronic acid and L-guluronic acid polymer from the cell wall of algae. Alginates are useful in neutralizing the charge of haze components generally known as protective colloids. In protective colloids, one polar or charged compound is adsorbed to the surface of another, causing the overall complex to repel similar species resulting in suspension (20) Agar has the ability to disrupt the protective colloid complexes, and reduce haze.

Activated carbon adsorbents are used to modify the sensory character of juices, wines and spirits. The vast number of pores in each particle gives carbon extremely high internal porosity and surface area., typically from 500 to 2,000 m^2/g (25). The forces that hold adsorbed molecules to the carbon are mostly weak Van der Waal's forces, thus carbon attracts more nonpolar molecules. The micropores in carbon are so small that compounds much larger than flavonoid dimers would be excluded. Interestingly, activated charcoal has been reported to catalyze the oxidation of phenols to quinones and maybe ethanol to acetaldehyde (25).

The clay bentonites are clarifying agents commonly used for protein adsorption to stabilize white wines and juices or red and pink wines. These agents may indirectly bind phenols that have complexed with proteins, and they can also bind anthocyanins, resulting in color loss (12,16,26,27). Bentonite is a montmorillonite, and the most commonly used form in the United States is sodium bentonite (Wyoming clay) (20). The adsorption of proteins or other soluble cationic constituents is due primarily to the cation exchange action of this clay. (28). Proteins with isoelectric point values above wine pH will carry a net positive charge and should be readily exchanged onto bentonite and thus removed. It is also known that bentonite can prevent oxidative browning due to the removal of oxidative enzymes and by reducing the levels of metals (12)

To assess the effects of fining on specific phenols and classes of phenolics, multi-level fining trials were carried out using six fining agents (gelatin, egg albumen, agar, PVPP, bentonite, and carbon) on a red Merlot wine. Specific phenols were analyzed using one reverse-phase HPLC method for the abundant phenols, along with another method specifically designed to assess concentrations of resveratrol and corresponding glucosides (piceid). Wine fined with each fining agent at one level was also tested for the ability to inhibit the oxidation of human LDL. The objective of this investigation was to determine the effect of fining on the phenolic composition and to relate changes in the phenolic composition to changes in the antioxidant activity towards LDL.

Materials and Methods

Wine Samples. The wine was a commercially vinified Merlot *Vitis vinifera* varietal from the 1994 vintage, produced at the Stag's Leap Winery, and was received in bulk as unfined, unfiltered wines which had completed malolactic fermentation. The wine's general chemical characteristics are shown in Table I and indicate that it is a typical Californian red wine, but low in sulfur dioxide (29).

Fining and Preparation of Wines. The fining agent selection and the levels of each tested, was based on those commonly used in practice based on discussions with experienced winemakers and other experts. The spray dried egg albumen (Nulaid Foods, Ripon, California) had minimum of 92% total egg white solids, and contained triethyl citrate/sodium lauryl sulfate as a whipping aid. A 10% aqueous solution, equivalent to the albumen concentration in liquid egg whites, was prepared by stirring at room temperature until dissolved, then refrigerating for 24 hours prior to use. The gelatin was derived from pork or beef skins and bones in standard fine granular form, with a gel strength of 100 bloom (Cellulo Company, Fresno, California). A 3% aqueous solution was prepared by continuous mixing in a heated water bath (130-140°F, 55-60°C) until dissolved. The solution was cooled to room temperature prior to use. The agar source was KLEAR-MOR®, a proprietary fining agent blended from agar and inert dispersing materials, including silica (Cellulo Company). A 1% aqueous solution was prepared by slowly bringing the solution to the boiling point in a water bath, with continuous, gentle stirring until smooth and creamy. The hot dispersion was slowly added to the wine with continuous mixing. A 5% solution of Wyoming Bentonite, or (sodium montmorillonite,

Table I. Characteristics of the Merlot wine

Vintner	Stag's Leap Winery, Napa, CA
Appellation	Napa Valley
Variety	Merlot
pH	3.30
Titratable Acidity*	6.6 g/L
Volatile Acidity*	0.2 g/L
Free SO_2	0.07 mg/L
Total SO_2	11.5 mg/L
Alcohol (vol)	12.6%
Residual Sugar	2 g/L

*Titratable acidity is expressed in tartaric acid equivalents, volatile acidity is expressed in acetic acid equivalents

VITABEN®, Cellulo Company), was prepared by dissolving the bentonite in warm, distilled water. The solution of was allowed to swell for 48 hours prior to use. The synthetic fining material polyvinylpolypyrolidone, or PVPP, used for this experiment was Polyclar ®VT (Cellulo Company). Since PVPP is insoluble in water, continuous stirring was utilized when preparing both the 5% aqueous suspension and when adding the suspension to the wine samples. The activated carbon was of the deodorizing type, specifically DARCO®, (Cellulo Company). A 5% suspension in water was prepared by thorough mixing.

The fining agents were pipetted into a large test tubes containing 25.0 mL of wine. Small amounts of distilled water were added to some samples so that all of the wines had the same final volume. The samples were vortexed and the entire volume was added to 25 mL glass screw-top vials which had no headspace and were protected from light exposure. After overnight storage at 12°C, the samples were decanted and filtered through 0.45 μm poly-tetrafluoroethylene (PTFE) filter. Fining trials were carried out over several days and to estimate day to day variances, the control (unfined) wine was prepared on six different days.

Analysis of Phenolic Compounds. A Hewlett-Packard (Palo Alto, CA) Model 1090 HPLC System, was used to determine the levels of specific phenolic components. The HPLC system was equipped with a ternary solvent delivery system, a diode array UV-VIS detector, and HP ChemStation software for data collection and analysis. Full chromato-graphic traces were collected at 280, 520, 316, and 365 nm, and spectra were collected on peaks. The stationary phase was a Hewlett-Packard LiChrosphere C-18 column, 4mm x 250 mm, with 5 μM particle size packing. Operating conditions include an oven temperature of 40°C, injection volume of 25 μL, and flow rate of 0.5 mL/minute. The method was based on a previously published method for phenolic components in wine (30) and used the modified solvent gradient shown in Table II. Solvent A was 50 mM dihydrogen ammonium phosphate, adjusted to pH 2.6 with orthophosphoric acid. Solvent

B consisted of 20% Solvent A in HPLC-grade acetonitrile. Solvent C was 0.2 M orthophosphoric acid adjusted with hydrochloric acid to pH 1.5.

Table II. Solvent Gradient for the HPLC Analysis of Phenolic Compounds

Time (minutes)	Solvent A (%)	Solvent B (%)	Solvent C (%)
5	100	0	0
8	92	8	0
20	0	14	86
25	0	18	82
35	0	21.5	78.5
70	0	50	50
75	100	0	0
80	100	0	0

All HPLC injections were performed in duplicate. (+)- Catechin and (-)-epicatechin, 280 nm, and quercetin, 365 nm (Aldrich, Milwaukee, WI), caffeic acid, 316 nm(Sigma, St. Louis, MO), gallic acid (MCB Manufacturing Chemists, Cincinnati, OH) and malvidin-3-glucoside, 520 nm (Pfaltz & Bauer, Waterbury,CT) were used as external standards at the indicated wavelengths. Caftaric acid was purified in our laboratory by a previously described method (31). The quercetin glycoside is expressed in quercetin equivalents, and all anthocyanins in malvidin-3-glucoside equivalents.

Resveratrol and piceid (resveratrol glucoside) isomers were measured using a different HPLC method than for the other phenolic compounds (32). The stationary phase was a Superspher C-18 column (Merck, Rathway, NJ), 4mm x 250 mm, with 5 μM particle size packing. Operating conditions include an oven temperature of 40°C, injection volume of 25 μL, and flow rate of 0.5 mL/minute. Solvent A was glacial acetic acid in water at pH 2.4 and solvent B was 20% solvent A in acetonitrile. The forty minute method began using 12% solvent B which gradually increased to 31.5% at 31 minutes followed by a column wash with 100% solvent B. Two wavelengths were monitored (286 and 306nm) and full spectra were collected on the peaks. Compounds were identified by comparing the retention times and UV spectra with standard compounds. Trans-resveratrol was obtained from Sigma and a piceid extract was obtained from *Polygonum cuspidatum* as previously described (33). The *cis* isomers of the aglycone and glycoside were obtained by light exposure to the *trans* isomers. Quantitation of *cis* and *trans* resveratrol were performed using a calibration curve for *trans* resveratrol (0-10mg/L). However, because the two isomers of resveratrol have different molar absorbtivities (34), corrections were applied. *Cis* and *trans* piceids were reported in *cis* and *trans* resveratrol equivalents, respectively.

To determine if the levels of the specific phenolics were significantly different from the unfined wine, the Dunnett's one-tailed t-test was performed using Statistical Analysis Software (SAS) (SAS Institute, Cary, NC) Values were considered significantly different if $p < 0.05$.

The Folin-Ciocalteau analysis of total phenols was performed using the method of Singleton and Rossi (35). This procedure used Folin & Ciocalteau's reagent (Sigma) and a saturated aqueous sodium carbonate solution to produce an absorbance at 765 nm. Gallic acid was used to make the standard curve and results are expressed as gallic acid equivalents (GAE).

Oxidation of Low Density Lipoprotein (LDL). Blood was drawn in ethylenediamine tetraacetic acid (EDTA) containing Vacutainer tubes (Becton Dickinson, Franklin Lakes, NJ) from five healthy volunteers. Low density lipoprotein (LDL) was isolated from the plasma by sequential density ultra-centrifugation (36). EDTA was removed by dialysis using Spectra/Por membrane tubing (Spectrum Medical Industries, Inc., M.W. cut off = 12-14 kDa) into pH 7.4 phosphate buffered saline brought to $4°C$ and deoxygenated by purging with nitrogen gas. The LDL protein concentration was determined using a Lowry protein analysis kit (Sigma) and was diluted to 1 mg protein /mL LDL solution with the buffered saline.

Antioxidant activity for human LDL was tested for the wines fined at one concentration below the highest levels of fining agents used in this study. The wines were dealchoholized by rotary evaporation and reconstituted with water. The antioxidant activity for human LDL was determined by the Frankel method (37). The wines were added to 250 µL of the LDL solution so the final phenol concentration in GAE was 5.0, 10.0, and 20.0 µM. The reaction was catalyzed by copper sulfate (80 µM, Fisher) and the vials were sealed with PTFE lined crimp caps and incubated for two hours in a $37°C$ shaking water bath. Inhibition of LDL oxidation was determined by monitoring hexanal production (Aldrich) by static headspace gas chromatography. A Perkin- Elmer (Norwalk, CT) 3B gas chromatograph (GC) equipped with an H-6 headspace injection port (Perkin Elmer), a capillary DB-1701 column (30m x 32µm, 1 µM thickness, J&W Scientific, Folsom, CA.) The oven, injector, and flame ionization detector were kept at $80°C$, $180°C$, and $200°C$ respectively. Immediately following the two hour incubation period, the vials were placed in the headspace injector, heated to $40°C$, pressurized with helium carrier gas for 30 seconds and a sample of the headspace was injected into the GC through the stationary injection needle. Results are expressed as % inhibition of the control LDL, i.e., $(C-S)/C*100$, where C was hexanal formed in the sample without any wine and S was hexanal formed in the sample with wine added. Replicate analysis were performed and results are expressed $\pm$ the standard deviation.

Results and Discussion

Total Phenolics by Folin Ciocalteau. The Folin-Ciocalteau total phenol level decreased with all fining agents and significantly with carbon and bentonite at all levels and PVPP at the highest level (Table III). One interference in the Folin-Ciocalteau assay is protein (29). This is of specific importance in this study because the wines are diluted to equivalent total phenol levels according to this assay for the subsequent LDL oxidation tests. New finished wines have very low levels of protein, however, the concentration of proteins that remain after fining has not been well documented.

HPLC Analysis. The HPLC method for phenolic compounds was capable of separating and quantifying monomeric (or small oligomeric such as dimeric phenolic compounds). The solvent gradient was modified slightly compared to previous work (30) to improve separations in congested areas of the chromatograms. In the replicate control samples, variation of components was 5-10% except for quercetin and epicatechin, which had higher variability (Table III). However, a large number of phenolic compounds in wine, in particular the oligomeric (4-10 monomeric units) and polymeric (condensed tannins) flavan-3-ols, are not separated by this reverse phase method, but in fact appear as a broad baseline drift during the separation (Figure 1). The components quantified were the major components with molecular weights less than approximately 1000 da, the unaccounted effects of fining agents on the undistinguished larger molecular weight components were an uncontrolled factor. The HPLC method used for separation of resveratrol and piceid was capable of complete separation of all four compounds in all of the wine samples. This method was very sensitive and had a limit of quantitation of 0.08mg/L for trans resveratrol. The coefficient of variation, determined from replicate injections over several days, was between 2 and 7% for all four compounds .

Specific Phenolic Compounds. The concentrations of caffeic and caftaric acids (hydroxycinnamates), and gallic acid (a benzoic acid) did not significantly differ from the control wine by the use of any of the fining agents at any of the levels tested.

The levels of catechin and epicatechin (monomeric flavan-3-ols) were not affected by fining with carbon, bentonite, or gelatin at any level. PVPP had the greatest effect on these compounds reducing levels of catechin and epicatechin to 60% and 74% of the control wine respectively at the highest treatment level. Additionally, agar and albumen also appeared to reduce the level of epicatechin, but the reduction of epicatechin was not statistically significant with the use of any of the fining agents.

Quercetin was not affected by gelatin or albumin fining at any of the levels tested, although the high variability in quercetin analysis limits the certainty of this statement. PVPP had the most dramatic effect on quercetin, reducing levels to 32% of the control, while bentonite reduced quercetin by approximately half at the highest level used. Carbon reduced the levels of quercetin to 60%, but the reduction was not statistically significant. The flavonol aglycones are less polar than most other flavonoids, and perhaps this property affected the interaction of quercetin with the fining agents. The levels of a quercetin glycoside changed far less dramatically than the aglycones which may be due to their increased polarity. This compound was reduced to approximately three fourths of the initial level by both PVPP and carbon.

Resveratrol and piceid isomers were decreased by all levels of carbon fining. At the highest level of carbon resveratrol isomers were reduced to about half of the initial levels while piceid isomers were reduced to 80% of the level found in the control wine. Similarly, the highest level of PVPP decreased the levels resveratrol isomers to 33% of the control wine while the piceid isomers were reduced to 80%. Conversely, gelatin did not significantly affect the levels of these compounds at any of the concentrations tested and albumen had only small effects on resveratrol and piceid. concentrations. Bentonite and agar fining also showed little effect on these compounds, even at the highest treatment levels.

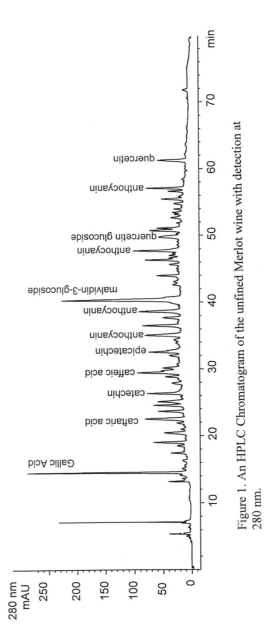

Figure 1. An HPLC Chromatogram of the unfined Merlot wine with detection at 280 nm.

Table III. The Effect of Fining agents on

Fining Agent Used	Fining Agent Level (mg/L)	Folin-Ciocalteau mg/L GAE)	Gallic Acid	Catechin	Epicatechin	t-Caftaric Acid	Caffeic Acid	Malvidin-3-Glucoside	A
Unfined Wine	0	2041 ± 101	31 ± 4	45 ± 6	42 ± 9	17 ± 2	3.0 ± 0.6	169 ± 15	
Carbon	240	1940 ± 109	28 ± 0	41 ± 0	40 ± 1	16 ± 0	3.3 ± 0.0	164 ± 1	
	480	**1817 ± 19**	28 ± 1	41 ± 0	41 ± 2	15 ± 0	3.2 ± 0.0	165 ± 1	
	960	**1737 ± 26**	28 ± 0	40 ± 1	38 ± 0	15 ± 0	3.2 ± 0.1	160 ± 0	
	1920	**1614 ± 49**	28 ± 0	40 ± 0	39 ± 0	15 ± 0	3.1 ± 0.1	157 ± 0	
Bentonite	120	**1839 ± 72**	28 ± 0	40 ± 1	41 ± 1	15 ± 0	3.3 ± 0.2	153 ± 4	
	240	**1852 ± 24**	30 ± 0	41 ± 0	42 ± 1	16 ± 0	3.4 ± 0.0	151 ± 0	
	480	**1852 ± 57**	31 ± 0	43 ± 0	42 ± 0	17 ± 0	3.5 ± 0.0	**138 ± 1**	
PVPP	60	2057 ± 23	33 ± 4	43 ± 4	43 ± 5	18 ± 2	3.7 ± 0.6	181 ± 21	
	120	2068 ± 76	32 ± 0	43 ± 2	43 ± 0	17 ± 0	3.8 ± 0.0	183 ± 1	
	240	1956 ± 36	28 ± 0	**31 ± 1**	36 ± 0	17 ± 0	3.3 ± 0.0	175 ± 3	
	480	**1809 ± 26**	26 ± 1	27 ± 0	31 ± 0	16 ± 1	2.9 ± 0.2	167 ± 2	
Gelatin	30	2089 ± 69	31 ± 3	45 ± 3	49 ± 3	18 ± 1	3.8 ± 0.3	169 ±15	
	60	2110 ± 55	32 ± 2	44 ± 3	49 ± 3	18 ± 1	3.9 ± 0.2	166 ±11	
	120	1967 ± 106	34 ± 2	47 ± 2	52 ± 1	19 ± 1	4.0 ± 0.1	180 ± 9	
	240	2013 ± 49	31 ± 1	44 ± 2	50 ± 2	17 ± 1	3.7 ± 0.0	167 ± 4	
Albumen	60	2040 ± 70	33 ± 2	41 ± 1	30 ± 2	17 ± 1	3.5 ± 0.0	165 ± 6	
	120	2051 ± 31	34 ± 1	44 ± 1	33 ± 1	18 ± 0	3.6 ± 0.2	174 ± 4	
	240	2007 ± 43	30 ± nd	40 ± nd	28 ± nd	16 ± nd	3.6 ± nd	152 ± nd	
Agar	120	2113 ± 42	30 ± 1	40 ± 0	30 ± 1	16 ± 0	3.5 ± 0.1	163 ± 3	
	240	2096 ± 28	29 ± 0	40 ± 0	30 ± 0	16 ± 0	3.6 ± 0.2	164 ± 1	
	480	2075 ± 53	28 ± 0	39 ± 0	29 ± 1	16 ± 0	3.5 ± 0.1	159 ± 2	

Bold values are significantly different at p < 0.05

Phenolic Components of Merlot Wine

Epicatechin	t-Caftaric Acid	Caffeic Acid	Malvidin-3-Glucoside	Other Anthocyanins	Quercetin	Quercetin Glycoside	Resveratrols	Piceids	Total by HPLC
42 ± 9	17 ± 2	3.0 ± 0.6	169 ± 15	222 ± 19	28 ± 10	47 ± 5	11.9 ± 0.3	9.4 ± 0.2	627 ± 49
40 ± 1	16 ± 0	3.3 ± 0.0	164 ± 1	215 ± 6	34 ± 4	43 ± 1	10.0 ± 0.1	8.8 ± 0.1	603 ± 11
41 ± 2	15 ± 0	3.2 ± 0.0	165 ± 1	216 ± 3	29 ± 2	43 ± 1	10.0 ± 0.1	8.5 ± 0.2	598 ± 5
38 ± 0	15 ± 0	3.2 ± 0.1	160 ± 0	204 ± 1	24 ± 1	40 ± 1	8.8 ± 0.1	8.0 ± 0.0	566 ± 1
39 ± 0	15 ± 0	3.1 ± 0.1	157 ± 0	192 ± 1	17 ± 0	36 ± 1	6.0 ± 0.2	7.4 ± 0.1	540 ± 1
41 ± 1	15 ± 0	3.3 ± 0.2	153 ± 4	206 ± 4	24 ± 8	42 ± 1	12.2 ± 0.0	9.9 ± 0.1	563 ± 29
42 ± 1	16 ± 0	3.4 ± 0.0	151 ± 0	205 ± 3	17 ± 2	44 ± 1	12.1 ± 0.1	9.8 ± 0.1	571 ± 5
42 ± 0	17 ± 0	3.5 ± 0.0	138 ± 1	188 ± 2	14 ± 1	44 ± 1	12.1 ± 0.4	9.9 ± 0.0	541 ± 7
43 ± 5	18 ± 2	3.7 ± 0.6	181 ± 21	234 ± 25	21 ± 4	49 ± 5	11.9 ± 0.0	9.1 ± 0.1	646 ± 70
43 ± 0	17 ± 0	3.8 ± 0.0	183 ± 1	240 ± 4	17 ± 1	48 ± 0	9.6 ± 0.4	8.8 ± 0.1	647 ± 7
36 ± 0	17 ± 0	3.3 ± 0.0	175 ± 3	213 ± 3	11 ± 3	42 ± 0	5.3 ± 0.1	8.0 ± 0.0	568 ± 2
31 ± 0	16 ± 1	2.9 ± 0.2	167 ± 2	199 ± 1	9 ± 2	37 ± 0	3.9 ± 0.1	7.5 ± 0.1	526 ± 3
49 ± 3	18 ± 1	3.8 ± 0.3	169 ± 15	231 ± 24	25 ± 2	47 ± 3	11.7 ± 0.2	9.1 ± 0.1	640 ± 55
49 ± 3	18 ± 1	3.9 ± 0.2	166 ± 11	226 ± 15	21 ± 4	46 ± 3	11.8 ± 0.2	9.1 ± 0.0	628 ± 42
52 ± 1	19 ± 1	4.0 ± 0.1	180 ± 9	241 ± 9	26 ± 0	48 ± 4	11.4 ± 0.0	9.1 ± 0.0	670 ± 28
50 ± 2	17 ± 1	3.7 ± 0.0	167 ± 4	225 ± 9	22 ± 1	45 ± 3	11.6 ± 0.2	9.2 ± 0.0	626 ± 21
30 ± 2	17 ± 1	3.5 ± 0.0	165 ± 6	206 ± 1	12 ± 2	46 ± 1	11.3 ± 0.1	8.8 ± 0.1	573 ± 10
33 ± 1	18 ± 0	3.6 ± 0.2	174 ± 4	227 ± 3	13 ± 1	48 ± 1	11.5 ± 0.1	8.4 ± 0.1	616 ± 9
28 ± nd	16 ± nd	3.6 ± nd	152 ± nd	197 ± nd	19 ± nd	43 ± nd	11.3 ± 0.1	8.2 ± 0.0	548 ± nd
30 ± 1	16 ± 0	3.5 ± 0.1	163 ± 3	213 ± 3	26 ± 0	45 ± 2	11.5 ± 0.1	9.0 ± 0.0	585 ± 10
30 ± 0	16 ± 0	3.6 ± 0.2	164 ± 1	216 ± 2	27 ± 2	44 ± 0	11.2 ± 0.0	9.2 ± 0.0	590 ± 5
29 ± 1	16 ± 0	3.5 ± 0.1	159 ± 2	211 ± 0	24 ± 1	43 ± 2	11.2 ± 0.0	9.2 ± 0.1	572 ± 6

Lastly, monomeric anthocyanins were not significantly affected by the use of any of the fining agents except carbon and bentonite at the highest treatment levels. However, even at the highest level of carbon, malvidin-3-glucoside was not decreased by more than 10%. The fining agent that caused the most dramatic decrease anthocyanin levels was bentonite and this fining agent reduced anthocyanins to 82% of the control at the highest treatment level.

Antioxidant Activity towards Human LDL. At equivalent total phenol levels (by the Folin Ciocalteau method), all samples had similar activity at 10.0 μM where inhibition was generally above 95%. At 20 μM GAE all wines inhibited more than 99% of LDL oxidation regardless of the fining agent used. However, the wines had very different antioxidant activities at 5.0 μM (Table IV). The antioxidant activity at 5.0 μM GAE increased with all fining agents and dramatically with PVPP and bentonite. Because the antioxidant activities were compared after dilutions to equivalent total phenol levels, it is important to note that the data shows which samples of phenolics have the most potency at the same concentration, not which sample has the highest antioxidant concentration.

Interestingly, the fining agent that mot effectively removed the monomeric phenols (PVPP), had the most dramatic increase in antioxidant activity at the same concentration. One explanation for the increase in antioxidant activity of this wine, is that, on the basis of total phenols (by the Folin Ciocalteau assay), the larger phenolic compounds (i.e. oligomers to polymeric condensed tannins) have superior antioxidant activity compared to the small molecular weight compounds. The larger compounds may be better chelators, and since this oxidation test is catalyzed by copper ions, such an effect should be significant.

A reason for the greater enhancement of antioxidant activity with PVPP treatment may be that the protein treatments may be leaving behind significant amounts of protein which is causing an increase in the apparent phenol concentration. Thus the apparent phenol concentration is higher than the true level, and thus lower levels of activity would be expected in the antioxidant tests. While the levels of residual protein in these treatments has not been described very well, there are reports that residual protein from such treatments can induce allergic reaction in those sensitive to related foods, i.e. albumin treated wines cause reactions in those allergic to eggs (38).

Bentonite fining also significantly increased the ability of the wine to inhibit LDL oxidation. Bentonite is used for protein adsorption but may indirectly bind phenols that have complexed with proteins. The protein content, and hence the removal of protein from this wine was not determined, however the protein content of most finished wines is usually low compared to the phenolic content. It is apparent from the Folin-Ciocalteau assay that this fining agent removed significantly more compounds than most of the other fining agents used in this study. The increased ability to inhibit LDL oxidation may be due to the removal of protein-tannin complexes, active in the Folin-Ciocalteau assay, that are not active inhibitors of LDL oxidation. Additionally, bentonite removed more phenolic compounds (mostly anthocyanins and quercetin) than gelatin, albumin, agar and carbon.

Conclusions

Confirming previous studies, PVPP was the most effective fining agent in removing smaller phenolic compounds. However, in this study, bentonite and carbon were also

Table IV. Percent Inhibition of Oxidation of Human Low-Density Lipoprotein by Merlot Wines Fined with Six Fining Agents [a]

Fining Agent Used	Fining Agent Level (mg/L)	% Inhibition 5.0 µM (GAE)	% Inhibition 10.0 µM (GAE)	% Inhibition 20.0 µM (GAE)
Control	na	26 ± 1	98 ± 1	99 ± 1
Agar	240	34 ± 6	95 ± 1	99 ± 1
PVPP	240	**51** ± 3	94 ± 1	99 ± 1
Bentonite	240	**40** ± 6	99 ± 1	99 ± 1
Albumen	240	35 ±11	96 ± 2	99 ± 1
Gelatin	120	26 ± 3	95 ± 1	100 ± 1
Carbon	960	31 ± 2	97 ± 1	99 ± 1

[a] Bold values are significantly different from the control p< 0.05

observed to have significant effects. PVPP had the most dramatic effect on quercetin, resveratrol and their glycosides and bentonite had the largest affect on anthocyanins. Carbon had significant effects on anthocyanins, flavonols and the stilbenes, but was not the strongest fining agent for any of the classes of phenolics over the levels tested, despite its reputation for effective fining. Albumin, gelatin and agar had very little effect on the levels of any of the monomeric phenolic compounds.

All of the fining agents removed components that responded to the Folin Ciocalteau assay, however the removal of these components did not decrease the antioxidant activity of the wines when normalized to equivalent total phenol levels. The wines fined with agents that removed the most monomeric phenolics, PVPP and bentonite, were the most potent inhibitors of LDL oxidation at the same phenolic concentration. We conclude that the inhibition of LDL oxidation *in-vitro* was significantly affected by fining with bentonite and PVPP, and was correlated with the reduction of monomeric phenolics, however, the effect of fining on wine components such as oligomeric or polymeric phenols and proteins may have been confounding factors in this assay.

Acknowledgments
We would like to thank the Stephen Sinclair Scott foundation and The Wine Spectator for financial support during this project. We would also like to thank Roger Boulton for his expert advice on wine fining and Edwin Frankel and Debra Pearson for their help with the oxidation experiments.

Literature Cited

1. Yokotsuka, K.; Singleton, V.L. *Am. J. Enol. Vitic.* **1987**, 38 (3), 199-206.
2. Baldwin, G. *Australian Grapegrower & Winemaker* **1992**, 344, 21-22.

3. Noble, A.C., In *Bitterness in Foods and Beverages. Developments in Food Science 25*; Rouseff, R.L., Ed.; Elsevier: New York, 1990; pp. 145-158; N.

4. German, J.B.; Frankel, E.N.; Waterhouse, A.L.; Hansen, R.J.; Walzem, R.L., In *Wine: nutritional and therateutic benefits*; ACS Symposium Series 661; Watkins, T., Ed.; American Chemical Society: Washington DC, 1997; pp. 196-214.

5. Frankel, E.N.; Kanner, J.; German, J.B.; Parks, E.; Kinsella, J.E. *Lancet* **1993**, 341, 454-457.

6. Esterbauer, H.; Gebicki, J.; Puhl, H.; Jürgens, G. *Free Radical Res. Commun.* **1992**, 13, 341-390.

7. Steinberg, D. *Journal of Internal Medicine* **1993**, 233, 227-232.

8. Demrow, H.S.; Slane, P.R.; Folts, J.D. *Circulation* **1995**, 91, 1182-1188.

9. Hagerman, A.E.; Butler, L.G. *J. Biol. Chem.* **1981**, 256 (9), 4494-4497.

10. Rossi, J.A., Jr.; Singleton, V.L. *Am. J. Enol. Vitic.* **1966**, 17, 240-246.

11. Van Buren, J.P.; Robinson, W.B. *J. Agric. Food Chem.* **1969**, 17 (4), 772-777.

12. Singleton, V.L. *Wines and Vines* **1967**, 48 (2), 23-26.

13. Slinkard, K.; Singleton, V.L. *Am. J. Enol. Vitic.* **1977**, 28 (1), 49-55.

14. Lea, A.G.H.; Timberlake, C.F. *J. Sci. Food Agric.* **1978**, 29 (5), 484-492.

15. Ricardo-da-Silva, J.M.; Cheynier, V.; Souquet, J.M.; Moutounet, M.; Cabanis, J.C.; Bourzeix, M. *J. Sci. Food Agric.* **1991**, 57 (1), 111-125.

16. Bravo Haro, S.; Rivas Gonzalo, J.C.; Santos Buelga, C. *Rev. Agroqui.* **1991**, 31 (4), 584-590.

17. Rapp, A.; Bachmann, O.; Steffan, H. *Bulletin de l'O.I.V.* **1977**, 553, 167-196.

18. Asano, K.; Shinagawa, K.; Hashimoto, N. *Am. Soc. Brew. Chem.* **1982**, 40 (4), 147-154.

19. Peri, C.; Bonini, V. *Journal of Food Technology* **1976**, 11 (3), 283-295.

20. Boulton, R.B.; Singleton, V.L.; Bisson, L.F.; Kunkee, R.E. *Principles and Practices of Winemaking*; Chapman & Hall: New York, New York, 1995.

21. Zoecklein, B.W.; Fugelsang, K.C.; Gump, B.H.; Nury, F.S. *Production Wine Analysis*; Chapman & Hall: New York, New York, 1989.

22. Singleton, V.L.; Noble, A.C. *Phenolic, Sulfur, and Nitrogen Compounds in Food Flavors, American Chemical Society Symposium Series* **1976**, 26, 47-70; N.

23. Cantarelli, C.; Giovanelli, G.; Gallizia, S. *Industrie delle Bevande* **1989**, 18 (101), 177-182.

24. Ough, C.S.; Amerine, M.A. *Am. J. Enol. Vitic.* **1960**, 11, 5-14.

25. Singleton, V.L. *Wines and Vines* **1964**, 45 (3), 29-31.

26. Dumazert, G. *Ind. Aliment. Agricol.* **1974**, 91 (1), 3-9.

27. Kovac, V.; Alonoso, E.; Bourzeix, M.; Revilla, E. *J. Agric. Food Chem.* **1992**, 40, 1953-1957.

28. Rodriguez, J.L.P.; Weiss, A.; Lagaly, G. *Clay and Clay Minerals* **1977**, 25, 243-251.

29. Ough, C.S.; Amerine, M.A. *Methods for analysis of musts and wines*, 2nd ed.; Wiley-Interscience: New York, 1988.

30. Sanduja, R.; Linz, G.S.; Alam, M.; Weinheimer, A.J.; Martin, G.E. *J. Heterocyclic Chem.* **1986**, 23, 529-535.

31. Singleton, V.L.; Salgues, M.; Zaya, J.; Trousdale, E. *Am. J. Enol. Vitic.* **1985**, 36, 50-56.

32. Lamuela-Raventós, R.M.; Romero-Pérez, A.I.; Waterhouse, A.L.; de la Torre-Boronat, M.C. *J. Agric. Food Chem.* **1995**, 42, 281-283.

33. Lamuela-Raventos, R.M.; Waterhouse, A.L. *Phytochemistry* **1994**, 37, 571-573.

34. Trela, B.C.; Waterhouse, A.L. *J. Agric. Food Chem.* **1996**, 44, 1253-1257.

35. Singleton, V.L.; Rossi, J.A. *Am. J. Enol. Vitic.* **1965**, 16, 144-158.

36. Orr, J.R.; Adamson, G.L.; Lindgren, F.T. *Analysis of Fats Oils and Lipoproteins*; American Oil Chemists' Society: Champain,Il, 1991.

37. Frankel, E.N.; German, J.B.; Davis, P.A. *Lipids* **1992**, 27, 1047-1051.

38. Marinkovich, V.A., In *Wine in Context: Nutrition, Physiology, Policy*; Waterhouse, A.L.; Rantz, J.M., Eds.; American Society of Enology & Viticulture: Davis, CA, 1996; pp. 26-28.

Chapter 12

Why Do Wines Taste Bitter and Feel Astringent?

Ann C. Noble[1]

Department of Viticulture and Enology, University of California, Davis, CA 95616

Bitterness and astringency in wine are elicited primarily by flavonoid phenols. Monomeric flavonoid phenols are primarily bitter, but upon polymerization astringency increases more rapidly than bitterness. Molecular conformation affects sensory properties: (-)-epicatechin is more astringent and bitter than its chiral isomer (+) catechin. In wine, perception of both bitterness and astringency are also affected by other wine components. Increasing viscosity or raising pH results in a decrease in perceived intensity of the tactile sensation of astringency, whereas little or no effect on intensity of bitter taste is observed. Higher concentrations of ethanol enhance bitterness intensity in wines, but have no effect on perception of astringency. Although the ability to taste propylthiouracil has not been shown to affect perception of wine, salivary flow status does affect perception of bitterness and astringency in wine. Subjects with high salivary flow rates perceived maximum intensity sooner and reported shorter duration of both bitterness and astringency than low-flow judges.

Red wines are characterized by bitterness and astringency, whereas white wines occasionally are bitter but seldom are astringent. In wine, both attributes are primarily elicited by the flavonoid polyphenolic compounds, which have been reviewed extensively elsewhere (*1- 3*). In this chapter, attention will be focused on the most recent investigations of factors which influence perception of bitterness and astringency.

Astringency is a tactile sensation, which is often described as a puckering, rough or drying mouthfeel. The mechanism of its perception is unknown, although it is probably mediated by touch or mechanoreceptors (*4*). Chemically, astringents have been defined as compounds which precipitate proteins. For water soluble phenols, this has been reported to require molecular weights between 500 and 3000 daltons (*5*).

[1]Telephone: 530–752–0387; Fax: 530–752–0382; email: ACNOBLE@UCDAVIS.edu.

Bitter taste is elicited by structurally diverse compounds, including phenols, ions, amino acids and peptides, alkaloids, acylated sugars, glycosides, nitrogenous compounds, and thiocarbamates. Taste receptor cells are primarily associated with papillae on the tongue. The signal transduction mechanisms by which taste perception occurs are well not understood, but are the focus of intensive research as reviewed recently (6).

Astringency and bitterness are both very persistent or lingering sensations, thus to quantify and characterize them most fully, temporal methods have been employed. From these continuous evaluations of perceived intensity over time, typical time-intensity (T-I) curves are developed from which several parameters can be extracted, such as time to maximum intensity, maximum intensity, total duration, decay rate etc. Time to maximum typically varies with the nature of the sensation being rated, but not with compound concentration. In contrast, as concentration of a compound is raised, the intensity at maximum and the total duration increase and are highly correlated (7). This can be seen in Figure 1, where the average intensity curves for bitterness over time are displayed for two wines varying in ethanol concentration. An additional complication in the study of bitterness and astringency, is the increase in perceived intensity on repeated ingestion. For example, Guinard *et al.* (8) demonstrated that perceived intensity of astringency increased when red wines were sipped at 20 sec intervals, whereas no increase was found with sipping at 40 sec intervals. Thus using the TI method by which intensity is rated continuously from ingestion through swallowing (or spitting) until sensation extinction provides a way to study these phenomena without the confounding effect of carry-over and buildup.

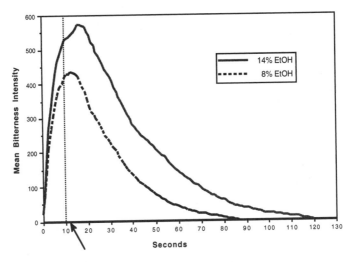

Figure 1. Mean time-intensity curves for bitterness of white wine with 8 and 14 % (v/v) ethanol (n=24 judges x 2 reps. The arrow denotes expectoration at 10 sec. (9).

Role of Phenolic Compounds.

Previous research has demonstrated that size of polyphenolic compounds affects their relative bitterness and astringency. Monomers of flavonoid and nonflavonoid phenolics are more bitter than astringent, whereas their polymers are more astringent than bitter (*10-12*). The relative duration of bitterness and astringency in white wines to which 1500 mg/L of catechin or tannic acid has been added illustrates this (see Figure 2). The flavan-3-ol monomer, catechin, increased bitterness and astringency total duration (Figure 2) and maximum intensity (not shown) over that of base wine, but the increase in astringency parameters was far less than that of bitterness. In contrast, a nonflavonoid polymer, tannic acid, produced a large increase in duration (Figure 2) and maximum intensity (not shown) of astringency and a smaller increase in the bitterness parameters (*9, 13*).

That larger molecules feel more astringent than smaller molecules is consistent with chemical measures of astringency: relative chemical astringency of flavonoid phenols, defined by the ability to precipitate protein, increases with molecular weight, from dimers to higher oligomers (*14*). Although Bate-Smith reported a minimum MW of 500 as a requirement for astringency, monomers, (+) catechin and (-) epicatechin (MW= 290) elicit astringency (*11, 12, 15*). Possibly the astringency elicited by the flavan-3-ol monomers is the result of unprecipitated protein-catechin complexes as reported by Yokotsuka and Singleton (*16*).

More recently benzoic acid derivatives (MW 122-170) have also been shown to be astringent (*17*). The most astringent compounds, salicylic (2-hydroxy benzoic acid) and gentisic (2,5 dihydroxy benzoic acid) acids, were *ortho* substituted, but neither had vicinal hydroxyl groups. Both derivatives had lower pHs than the non-*ortho* substituted ones, which may have contributed not only to sourness but also to astringency. McManus *et al* (1981) (*18*) proposed previously that simple phenols which contain 1,2 dihydroxy or 1,2,3 trihydroxy groups (such as epicatechin or catechin) may cross link and thereby precipitate proteins. It could be speculated that *ortho* substitution conveys some kind of binding capability similar to that of flavan-3-ols or polyphenolics of higher MW.

Small differences in configuration, such as that between between chiral isomers (+) catechin (Cat) and (-) epicatechin (Epi) (Figure 3) confer differences in sensory properties: epicatechin is more bitter and astringent than catechin (*15,19*). The conformation of the C ring of epicatechin is more planar than catechin, possibly causing its higher astringency due to the greater availability for intermolecular hydrogen bonding of the hydroxyl group in the 3 position (*15*). Since the relationship between molecular structure and bitterness has not been identified, no speculation about the configurational differences on bitterness is made.

Astringency and bitterness of the monomers (Cat and Epi), three dimers and two trimers, synthesized from catechin and epicatechin by condensation of the monomeric procyanidins with (+)-dihydroquercitin, were evaluated by T-I. Consistent with the previous studies cited above, as the degree of polymerization increased, perceived bitterness Imax and Ttot decreased whereas astringency Imax increased (Table I). The bond linking the monomeric units also had an influence on the sensory properties. The catechin-catechin dimer (Cat-Cat) linked by a 4-6 bond was more bitter than the Cat-Cat 4-8 dimer and the catechin-epicatechin 4-8 dimer. However, astringency appeared to vary as a function of the identity of the monomeric units and the site of the linkage with the Cat-Cat 4-8 dimer being least astringent (*19*).

BITTERNESS

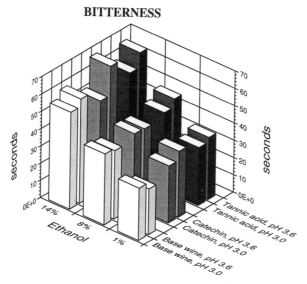

ASTRINGENCY

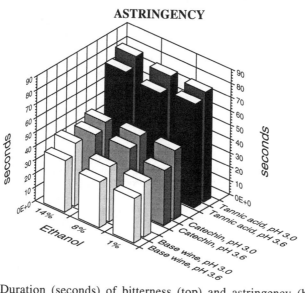

Figure 2. Duration (seconds) of bitterness (top) and astringency (bottom) as a function of ethanol, phenolic composition, and pH (n=48) LSD=5.24 (bitterness) and 6.53 seconds (astringency) (9)

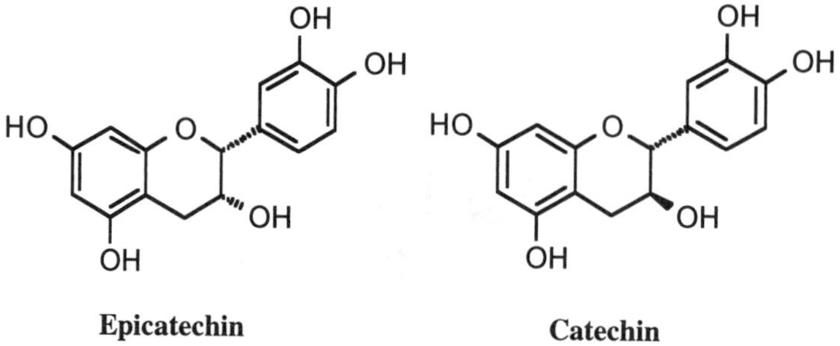

Epicatechin   **Catechin**

Figure 3. Structures of 3R(-)-epicatechin and 3S (+)-catechin.

Table I. Maximum Intensity (MAX) andTotal Duration (TOT) of Bitterness and Astringency of Flavan-3-ol monomers, dimers and trimers (n= 18 judges x 2 reps) (19)

| | **BITTERNESS** | | **ASTRINGENCY** | |
Compound	**MAX**	**TOT (s)**	**MAX**	**TOT (s)**
(-) Epicatechin	508	30	325	29
(+) Catechin	456	30	300	27
Catechin(4->8)Catechin	335	26	340	28
Catechin(4->8)Epicatechin	373	26	413	30
Catechin(4->6)Catechin	477	32	431	32
Catechin(4->8)Catechin(4->8)Catechin	244	24	376	26
Catechin(4->8)Catechin(4->8)Epicatechin	275	24	428	34
LSD (0.05)	80.3	3.0	71.3	5.2

Role of Ethanol

In addition to phenolic compounds, other factors in wine which elicit or enhance bitterness and astringency include: ethanol, sugar, and organic acids. As illustrated in Figure 1, the bitterness intensity is higher and duration persits longer for a wine with higher ethanol concentration. Bitterness intensity (not shown) was greater and persistence was extended longer (Figure 2) by an increase of 6% (v/v) in the ethanol concentration than by addition of 1500 mg/L catechin or tannic acid (9). In contrast varying the concentration of ethanol had very little effect on astringency (9) or sourness (20).

Role of Acid

Lowering the pH of the wines had a small and inconsistent effect on bitterness duration, but significantly increased duration of astringency (Figure 2). Fischer (9) proposed that this enhancement of astringency at lower pH may be explained by the higher ratio of charged phenolate ions at higher pHs. Although the pK_a of the phenolic hydroxyl group in flavonoids is 9.9 (21), pH differences below 4.0 will alter

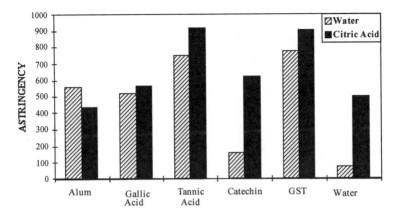

Figure 4: Mean astringency intensity ratings of astringents in water and in citric acid (1.5 g/L) (n = 18 judges x 2 reps; LSD = 104.54) (*24*).

the abundance of charged phenolate ions. Since the charged species are unable to participate in hydrogen bonding, and hence reduce interaction of the phenolic compounds with proteins, this may contribute to the decrease in perception of astringency at lower pHs.

Fischer's data are consistent with the increase in astringency intensity in red wine reported upon addition of tartaric acid (*22*) and malic or lactic acid (*23*). Solutions of citric acid and selected phenolic compounds were higher in astringency than the corresponding unacidified solutions of phenolic compounds in water (Peleg, H, Bodine, K., and Noble, A.C. submitted) As illustrated in Figure 4, in contrast to the enhancement of astringency of phenolic compounds upon acid addition, astringency of alum decreased. Alum is not found in wine but is is widely used in psychophysical research as an astringent stimulus (*24*). It should be noted that citric acid alone elicits a strong intensity of astringency. Both organic and mineral acids were first reported to contribute astringency by McDaniel and her coworkers (*25, 26*).

Role of Acid vs pH

The astringency of organic acids was correlated to pH by Lawless *et al* (*27*), and shown recently to be solely a function of pH, and not of the anion or titratable acidity (*28*). When equinormal acid solutions at three pH levels and at three normality levels at constant pH were rated for intensity of sourness and astringency, astringency increased only with an increase in hydrogen ion concentration (decrease in pH). Increasing the normality and correspondingly the potential hydrogen ions (titratable acidity) had no effect on astringency perception, although increasing normality or decreasing pH produced the expected increases in sourness (*28*). The same astringency responses shown in Figure 5 for citric acid were also found for tartaric, malic, and lactic acids (*28*). Astringency elicted by aqueous solutions of acids perhaps is the result of precipitation of salivary proteins, but has not been investigated.

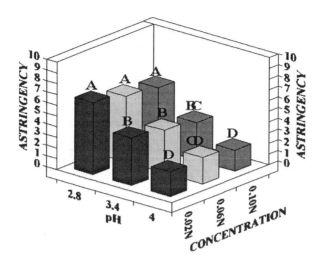

Figure 5: Mean astringency intensity of citric acid solutions. Means with same superscript do not differ significantly (p<0.05)(n = 14 judges x 2 reps); LSD=1.11 (28).

Role of Viscosity vs Sweetness

Addition of sucrose to red wine decreased astringency maximum intensity and total duration (29). To examine the separate effects of viscosity and sweetness on astringency, aqueous solutions of grape seed tannin thickened with carboxymethyl cellulose or sweetened with the non-carbohydrate sweetener, aspartame were evaluated in time-intensity studies. Increasing viscosity had no effect on bitterness temporal parameters, but significantly reduced astringency intensity. Conversely increasing sweetness did not affect astringency (Figure 6) although it reduced bitterness (30).

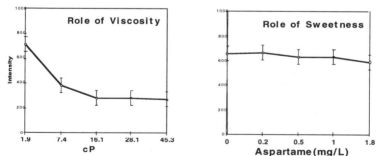

Figure 6. Effect of viscosity (cp) (left) and sweetness (right) on maximum intensity of astringency (31).

Role of Physiological Factors.

PROP status. Individuals classified as tasters of the bitter compound propylthiouracil (PROP) (not found in wine) have been reported to perceive bitterness more intensely and have a higher number of taste pores per taste bud and higher density of fungiform taste papillae on the tongue than non-tasters of PROP (*32- 36*). Despite this, PROP status has not been demonstrated to affect perception of bitterness or astringency of phenolic compounds in wine (*13, 29*) or water (*15, 17, 19, 30*).

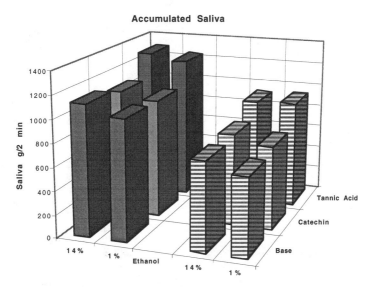

Figure 7. Parotid saliva flow (g) accumulated over two min in response to wines varying in added Tannic acid (1500 mg/L), catechin (1500 mg/L), ethanol (%v/v) and in pH. Solid bars pH 3.0 and striped bars pH 3.6 (n= 11 Judges x 2 reps) (*9*).

Salivary-flow status. Saliva contains proline-rich proteins (PRPs), of which the basic PRPs have been shown to have a high affinity for binding polyphenols (*37*). Unilaterally monitoring salivary flow of the parotid salivary gland in response to a subset of the wines shown in Figure 2, revealed increasing acid (lowering pH) or tannic acid stimulates salivary flow (*9*). The most astringent wines (which were spiked with tannic acid) and those at the lower pH (solid bars) elicted significantly higher output of saliva than wines without the tannic acid or at the higher pH (striped bars) (Figure 7). The intensely bitter 14 % ethanol wines elicited only slightly higher flow than the wines with only 1% ethanol, suggesting that neither ethanol nor perception of bitterness have a major effect on salivary production.

When the judges were partitioned into groups on the basis of their salivary flow rates, the high-flow subjects differed in their perception of both bitterness and astringency from the low-flow subjects. High-flow subjects perceived maximum intensity later, less intensely and for a shorter time than the low-flow subjects (*13*). The same comparative difference between high- and low-flow judges was found for perception of astringency in red wines as shown in Figure 8.

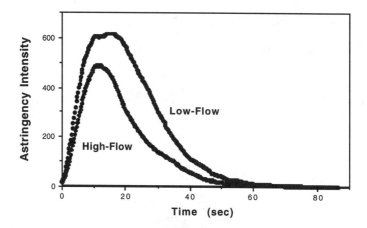

Figure 8. Mean time-intensity curves for astringency for 9 low-flow and 9 high-flow judges pooled across 10 red wines (Reproduced with permission from reference 29. Copyright 1995.)

Hypothesis for Mechanism of Astringency

The oral sensation of astringency elicited by polyphenolic compounds presumably is linked to precipitation of salivary proteins, and most effectively the basic PRPs. If precipitation of salivary proteins reduces the effectiveness of saliva as an oral lubricant, the "rough" feeling or astringent sensation may well be the result of increased perception of friction between oral surfaces as the ability of saliva to serve as an effective lubricant is decreased. Given that astringency is decreased as viscosity is increased, it may be speculated that raising viscosity restores lubrication in the oral cavity. High-flow individuals, who produce a higher total amount of salivary protein (38), may perceive less astringency because of their greater ability to restore lubrication. Experiments to measure the change in lubrication of protein solutions upon addition of phenolic compounds are in progress to explore this speculation.

Conclusion

The bitter taste and astringent feel of wines result primarily from phenolic compounds. The relative intensity of bitterness and astringency of phenolic compounds varies with the molecular size and conformation. Organic acids in wine, in addition to causing sourness increase the astringency of wine, but do not affect bitterness. In contrast, raising the ethanol level of table wines, significantly increases the intensity of bitterness, but does not affect astringency.

Acknowledgments

The contribution of K. Bodine, U. Fischer, K. Gacon, T. Ishikawa, H. Peleg, J. Robichaud, A. Smith, R. Sowalsky, and J. Thorngate and partial funding by the American Vineyard Foundation and BARD are gratefully acknowledged.

Literature Cited.

1. Singleton, V. L.; Noble, A. C. In *Phenolic, Sulfur, and Nitrogen Compounds in Food Flavors*; Charalambous, G., Ed.; Am. Chem. Soc.: Washington, D.C.; **1976**; pp. 47-70.
2. Noble, A. C. In *Bitterness in Foods and Beverages. Developments in Food Science.*; Rousseff, R. L., Ed.; Elsevier Science: N.Y.; **1990**; pp. 145-158.
3. Noble, A. C. *Physiol Behav* **1994,** 56 (6), 1251-1255.
4. Martin, J. H.; Jessell, T. M. In *Principles of Neural Science*;,Kandel, E.R;Schwartz, J. H.;Jessell, T. M., Ed.; Appleton & Lange: Norwalk, CT; **1991**; pp. 341-352.
5. Bate-Smith, E. C.; Swain, T. In *Comparative Biochemistry III, Constituents of life, Part A*; Mason, H. S.;Florkin, A. M., Ed.; Academic Press: N.Y.; **1962**; pp. 755-809.
6. Kinnamon, S. C. *Food Qual Pref* **1996,** 7, 153-159.
7. Noble, A. C. *Am J Enol Vitic* **1995,** 46, 128 - 133.
8. Guinard, J.-X.; Pangborn, R. M.; Lewis, M. J. *Am. J. Enol. Vitic.* **1986,** 37, 184-189.
9. Fischer, U. **1990** . M.Sc. Thesis, Univ. California, Davis, CA.
10. Lea, A. G. H.; Arnold, G. M. *J Sci Food Agric* **1978,** 29, 478-483.
11. Arnold, R. A.; Noble, A. C.; Singleton, V. L. *J Agric Food Chem* **1980,** 28, 675-678.
12. Robichaud, J. L.; Noble, A. C. *J Sci Food Agric* **1990,** 53, 343-353.
13. Fischer, U.; Boulton, R. B.; Noble, A. C. *Food Qual Pref* **1994,** 5, 55-64.
14. Haslam, E. *Biochem J* **1974,** 139, 285-288.
15. Thorngate, J. H.; Noble, A. C. *J Sci Food Agric* **1995,** 67, 531-535.
16. Yokotsuka, K.; Singleton, V. L. *Am J Enol Vitic* **1987,** 38, 199-206.
17. Peleg, H.; Noble, A. C. *Chem Senses* **1995,** 20, 393-400.
18. McManus, J. P.; Davis, K. G.; Lilley, T. H.; Haslam, E. *J.C.S .Chem Comm* **1981,** 309-311.
19. Gacon, K.; Peleg, H.; Noble, A. C. *Food Qual Pref* **1996,** 7, 343-344.
20. Fischer, U.; Noble, A. C. *Am J Enol Vitic* **1994,** 45, 6 - 10.
21. Oh, H. I.; Hoff, J. E. *J Food Sci* **1987,** 52 (5), 1267-1271.
22. Guinard, J.-X.; Pangborn, R. M.; Lewis, M. J. *J Sci Food Agric* **1986,** 37, 811-817.
23. Kallithraka, S.; Bakker, J.; Clifford, M. N. *J Food Sci* **1997,** 62, 416-420.
24. Bodine, K. K. 1996 . M. Sc. Thesis, Univ. of California, Davis, CA.
25. Rubico, S. M.; McDaniel, M. R. *Chem Senses* **1992,** 17, 273-289.
26. Hartwig, P.; McDaniel, M. R. *J Food Sci* **1995,** 60, 384-388.
27. Lawless, H. T.; Horne, J.; Giasi, P. *Chem Senses* **1996,** 21, 397-403.
28. Sowalsky, R. A. 1996 . M. Sc. Thesis, Univ. California, Davis, CA.
29. Ishikawa, T.; Noble, A. C. *Food Qual Pref* **1995,** 6, 27-33.
30. Smith, A. K.; June, H.; Noble, A. C. *Food Qual Pref* **1996,** 7, 161-166.
31. Smith, A. K. 1996 . M.Sc. Thesis, University of California, Davis, CA.
32. Miller, I. J.; Reedy, F. E. *Physiol Behav* **1990,** 47, 1213-1219.
33. Miller Jr., I.; Bartoshuk, L. M. In *Smell and Taste in Health and Disease*; Getchell, T. V.;Bartoshuk, L. M.;Doty, R. L.;Snow Jr., J. B., Ed.; Raven Press: New York; **1991**; pp. 205 - 233.
34. Bartoshuk, L. M. *Food Qual and Pref* **1993,** 4, 21-32.
35. Bartoshuk, L. M. In *Molecular Basis of Smell and Taste Transduction*; Ciba Foundation.: England; **1993**; pp. 251-267.
36. Bartoshuk, L. M.; Duffy, V. B.; Miller, I. J. *Physiol Behav* **1994,** 56, 1165-71.
37. Hagerman, A. E.; Butler, L. G. *J Biol Chem* **1981,** 256, 4494-4497.
38. Froehlich, D. A.; Pangborn, R. M.; Whitaker, J. R. *Physiol Behav* **1987,** 41, 209-217.

Chapter 13

Characterization and Measurement of Aldehydes in Wine

Susan E. Ebeler and Reggie S. Spaulding

Department of Viticulture and Enology, University of California, Davis, CA 95616

Short chain, volatile aldehydes contribute important sensory properties to wines and can affect aging and color stability. Methods for measuring these compounds in grapes, musts, and wines have been available for some time, however, they can be time-consuming, non-specific, or result in artifact formation during analysis. Reactions with bisulfite and phenolics also complicate analyses.   A gas chromatographic procedure for analysis of volatile aldehydes has been developed where the aldehydes are reacted with cysteamine to form stable thiazolidine derivatives. We evaluated this procedure for the analysis of short-chain saturated aldehydes (formaldehyde, acetaldehyde, propionaldehyde, etc.) in white and red wines. The method provides the opportunity to monitor individual aldehyde levels in wine during fermentation and aging and to evaluate the effects that aldehyde formation may have on wine flavor.

Short chain, volatile aldehydes are important to the flavor of a number of foods and beverages, including wine, contributing flavor characteristics ranging from "apple-like" to "citrus-like" to "nutty" depending on the chemical structure (Table 1). In wine, acetaldehyde is generally the aldehyde present in highest concentrations. It has a reported sensory threshold of 100 - 125 mg/L (1) and is an important flavor constituent of sherry and aged wines. Guth and co-workers in a separate chapter of this volume have shown that acetaldehyde and isovaleraldehyde are also important odor impact compounds in Gewürztraminer and Scheurebe wines. However, the concentrations, flavor properties, and sensory thresholds of other aldehydes in wine and alcoholic beverages are largely unknown.

Aldehydes also affect the aging characteristics and color stability of wines. Reactions with SO_2 decrease the amount of free SO_2 available to act as an antioxidant during wine storage. Acetaldehyde in particular can catalyze the condensation of flavonoids to form polymeric pigments which directly affect the taste and color of red wines (2).

Finally, the aldehydes are highly reactive and can bind *in vivo* to biological nucleophiles such as proteins, DNA, cellular membranes, and enzymes, resulting in toxic, mutagenic, and carcinogenic effects (3-6) . Whether aldehydes consumed in foods and beverages exhibit significant absorption and reactivity *in vivo* is not clear.

Table 1. Flavor characteristics of volatile, short chain aldehydes.

Aldehyde	Flavor Characteristic[1]
Formaldehyde	Sharp, pungent odor
Acetaldehyde	Overripe bruised apples, nutty, sherry-like
Propanal	Similar to acetaldehyde
Butanal	Pungent
2-Methyl-1-propanal (Isobutanal)	Characteristic, slightly apple-like
Pentanal	Warm, slightly fruity, nut-like, pungent at high concentrations
3-Methyl-1-butanal (Isovaleraldehyde)	Warm, herbaceous, slightly fruity, nut-like, penetrating, acrid at high levels
2-Methyl-1-butanal	Cocoa, coffee-like, sweet, slightly fruity, powerful, choking at high levels
Hexanal	Green, grassy, fruity
Heptanal	Fatty, unpleasant
Octanal	Sharp, fatty, fruity
Nonanal	Fatty, orange-rose-like, citrus-like

[1]References: 1, 51

However, air exposure to short-chain aldehydes such as formaldehyde and acetaldehyde is known to result in eye, nose and throat irritation and burning; coughing; dermatitis; and pulmonary edema (6, 7).

For these reasons, characterization of the concentrations, formation, reactions, and sensory properties of aldehydes in foods and beverages, including wines, is important. Development of analytical techniques which can accurately measure these individual aldehydes is necessary to understand and ultimately control aldehyde formation during processing and storage of wines and other alcoholic beverages. This article focuses on the formation and reactions of aldehydes in wines and describes analytical methodologies used to determine aldehyde concentrations. Finally, application of a simple gas chromatographic procedure which measures saturated aldehydes as their thiazolidine derivatives is described.

Formation of Aldehydes in Grapes and Wines. Aldehydes, particularly the C6 aldehydes (hexanal, *cis-* and *trans*-2-hexenal, cis-3-hexenal), are formed from fatty acid precursors in the grapes by the activity of oxido-reductase enzymes (*e.g.*, lipoxygenase) following crushing or maceration (8). Joslin and Ough (9) observed rapid formation of high levels of hexanal and *trans*-2-hexenal following crushing of French Columbard grapes and leaves. They also observed that addition of 100 ppm SO_2 decreased the amounts of C6 compounds formed, possibly through enzyme inhibition. Once formed however, the aldehydes are rapidly reduced to the corresponding alcohols (9, 10). Isomerization of *cis*-3-hexenal to *trans*-2-hexenal also readily occurs, especially if samples are heated as occurs during sample preparation by steam distillation and analysis by gas chromatography (9). Increased skin contact time increases formation of the C6 aldehydes (11) while carbonic anaerobiosis has recently been shown to decrease the amounts of these compounds formed (12).

Aldehydes also arise as normal by-products of yeast fermentation. Acetaldehyde is the ultimate electron acceptor in the conversion of glucose to ethanol. In this pathway, aldehyde dehydrogenase (ADH) reduces acetaldehyde to ethanol with the corresponding oxidation of NADH. Acetaldehyde levels are therefore dependent on the fermentation conditions, *e.g.*, temperature, O_2 levels, pH, SO_2 levels, and yeast nutrient availability (13, 14). Yeast strain can also affect aldehyde formation and excretion (15-17). For example, film yeasts used in sherry production are selected for their ability to produce very high acetaldehyde levels (18).

Aldehydes, especially the longer chain saturated and branched chain aldehydes (i.e., propanal, butanal, 2-methyl-1-propanal, 2-methyl-1-butanal, and 3-methyl-1-butanal) are also intermediates in the formation of fusel oils. These pathways involve anabolic metabolism of sugars or transamination of amino acids. During ethanol fermentation, the aldehydes may be reduced to the corresponding alcohols by ADH enzymes and excreted into the media. Herraiz et al. (19) found that longer chain aldehydes were not as readily reduced and excreted by the yeast, e.g., 35% reduction was observed for pentanal compared to 3% reduction for decanal.

Chemical oxidation reactions and Strecker degradation of amino acids may also result in aldehyde formation (16). These reactions are important for the formation of acetaldehyde in baked or heated Sherries. Chemical oxidation reactions catalyzed by metal ions (Cu, Fe, Mn, Mo) can result in aldehyde production in oak aged distillates (20). Such reactions may also result in the formation of oxidation products from fatty acids released during the secondary fermentation of sparkling wines (21).

Finally, Wildenradt and Singleton (22), have proposed that aldehyde formation during the aging of wine is largely a result of coupled oxidation reactions with phenolics. In the presence of oxygen, vicinal di- and tri-hydroxyphenols (e.g., caffeic acid, catechin, myricetin) are oxidized to the quinone with the corresponding production of a strong oxidant, postulated to be hydrogen peroxide (H_2O_2). This oxidant (H_2O_2) is capable of readily oxidizing alcohols in the wine, including ethanol, resulting in production of 1 mole of aldehyde per mole of oxidizable phenol.

Chemical Reactions of Aldehydes in Grapes and Wines. The short chain, volatile aldehydes are quite reactive making accurate analysis and quantitation difficult. In addition, many of these reactions are equilibrium reactions with the concentration of unreacted and reacted aldehydes being highly dependent on the analysis conditions.

Bisulfite addition products are readily formed at wine pHs (1, 23, 24). The bisulfite addition product is thought to be a more "sensory-neutral" compound and may be exploited by winemakers as a means of decreasing the aldehydic character of wines (1). Bisulfite addition has also been used to mask the stale flavor of beer which is thought to be largely due to the formation of *trans*-2-nonenal (25). Kaneda et al. (25) used HPLC with fluorescent detection of an o-phthalaldehyde derivative to quantitate and identify individual aldehyde-bisulfite products, however, only acetaldehyde-bisulfite adducts were observed in commercial beers with this method. Hydrolysis of the adducts occurs at pHs greater than 8, therefore by adjusting the pH prior to analysis, total aldehydes (free plus bisulfite bound) can be estimated. At low pHs accurate estimation of free aldehydes is complicated however, by analysis conditions which alter the equilibrium between bound and free forms (temperature, dilution, solvent extraction, analysis time, etc.).

Aldehyde-tannin and aldehyde-anthocyanin condensation reactions result in polymer formation (Figure 1). These polymers may be responsible for haze formation in wine and the polymers may eventually precipitate out of solution (26). The polymerized tannins have different flavor properties than the monomeric starting units (27-29) and formation of anthocyanin polymers affects wine color. In addition, these reactions may result in a reduction of aldehyde flavors in the wine. These condensation reactions are discussed more fully in other chapters of this volume. The formation of strong covalent bonds between the aldehyde and the tannin or anthocyanin makes recovery of the bound aldehydes difficult.

Acetals are equilibrium products between aldehydes and alcohols. As discussed by Williams and Strauss (30) acetals generally have less intense aromas than the corresponding alcohols and aldehydes. 1,1,3-Triethoxypropane and diethoxybutan-2-one (derived from acrolein and diacetyl, respectively) are common acetals in the heads fractions from continuous stills; acetals from other aldehydes including acetaldehyde, propanal, isobutanal, and isovaleraldeyde are also common (30). The equilibrium between the aldehyde and the acetal is highly dependent on alcohol concentration and pH, again making accurate quantitation of either the aldehyde or the acetal dependent on the analytical conditions (*e.g.*, sample dilution, solvent extraction, etc.) (30).

Finally, aldehydes can react with nitrogen (31-32) and sulfur nucleophiles, including H_2S, which may also be present in wines. These reactions may have dramatic effects on flavor and aroma (*e.g.*, formation of ethyl mercaptan from acetaldehyde and H_2S results in formation of a "onion-like" or "burnt rubber" aroma) and will decrease the levels of "free" aldehydes which can be readily quantitated (1).

Measurement of Aldehydes in Grapes and Wines. Because total aldehyde levels can vary significantly depending on yeast strain, nutrient composition, and fermentation and storage conditions, and because their presence may have an important impact on overall wine quality, knowledge of the concentration of these compounds at various stages during wine making is important. However, most analytical methods measure only acetaldehyde or are non-specific giving only a combined measure of total aldehyde levels. As discussed above, analysis of these aldehydes in wines is also complicated by their ability to form complexes with a number of other wine components at the acid pH's normally encountered in wines.

Distillation techniques, commonly employed to measure total saturated aldehyde levels (free and bound), utilize elevated temperatures during the analysis which can result in aldehyde loss through volatilization or artifactual formation during the heat treatment (33). The standard titrimetric procedures measure aldehydes, following distillation, by titrating excess bisulfite that has not complexed with the aldehydes

(34). Because all aldehydes can bind with bisulfite, it is not possible to quantitate each aldehyde individually using these methods.

Colorometric procedures involving reaction of aldehydes with hydrazines, semicarbazide, or piperidine/nitroprusside solutions are also non-specific and lack sensitivity (15, 35, 36). Schmidt et al. (33) have proposed an HPLC method for analyzing the 2,4-dinitrophenylhydrazone (DNPH) derivatives of specific aldehydes. This procedure allows for a number of aldehydes to be separated and measured simultaneously, however, HPLC methods in general suffer from poor resolving power and may have low sensitivity (37). In addition, hydrazine derivatizations are often performed under acidic conditions for maximal reactivity; these conditions would not provide quantitative information on total aldehyde content.

Enzymatic methods have also been employed for measurement of acetaldehyde levels. The affinity of these enzymes for other aldehydes has not been completely determined (13, 38-40), therefore, these methods also suffer from lack of specificity.

Direct analysis by Gas Chromatography (GC) following distillation/extraction or headspace sampling has been proposed (1, 24, 41). Again, however, volatilization and high reactivity of the aldehydes makes accurate quantitation difficult using these techniques. Hobley and Pamment (42) have also observed that some, but not all, bound acetaldehyde complexes decompose in the injection port of the GC, complicating measurement of the free and bound aldehydes.

Finally, formation of O-(2,3,4,5,6-pentafluorobenzyl)-hydroxylamine (PFBOA) derivatives and analysis by GC-Mass Spectrometry (GC-MS) and GC-Electron-Capture Detection (GC-ECD) appears to be a promising technique. de Revel and Bertrand (42, 43) used PFBOA derivatization to analyze a number of saturated and unsaturated aldehydes in wines, however, high concentrations of acetaldehyde made accurate quantitation of the other aldehydes present in lower concentrations difficult, depending on the wine matrix; the aldehydes were not always well separated from other chromatographic peaks; pH conditions for the derivatization were not specified; and removal of excess PFBOA by acidification caused the partial loss of some aldehydes. In addition, no specific information regarding derivatization efficiency and recovery, or absolute limits of detection and quantitation were reported by these authors.

An ideal analytical method for measuring aldehydes in wine would allow a number of aldehydes to be measured specifically and sensitively in one assay. In addition, the method would give a measure of free aldehyde levels and those that are bound to SO_2 or phenolics. A gas chromatographic (GC) procedures for the analysis of volatile aldehydes has been developed in which the aldehydes are reacted with a derivatizing agent at neutral or slightly basic conditions to form a stable derivative which can be easily analyzed (44) (Figure 2). The derivatizations are conducted at room temperature and the stable derivatives are quantified using nitrogen-phosphorous (NPD) or flame photometric (FPD) detectors for increased sensitivity and selectivity. Using these procedures, a number of aldehydes can be measured simultaneously and their identifications can be confirmed by GC-MS. Yasuhara and Shibamoto (45, 46) used these procedures to simultaneously measure a series of saturated and branched chain aldehydes in model systems with a detection limit of 5.8 pg for formaldehyde. The methods have been utilized to measure aldehydes in coffee, pork fat, air, and biological samples (46-49) but not in alcoholic beverages.

Materials and Methods

Materials. All reagents were obtained from Aldrich Chemical Co (Milwaukee, WI). Purities were as follows: Formaldehyde, 37 wt %; Acetaldehyde, 99.5%; Butanal, 99%; Propanal, 97%; 2-Methyl-1-propanal, 99%; Pentanal, 99%; 3-Methyl-1-butanal, 97%; Hexanal, 98%; Heptanal, 95%; Octanal, 99%; Nonanal, 95%; 2,4,5-trimethylthiazole (IS), 98%, 2-aminoethanethiol (cysteamine), 98%. Chloroform (Optima Grade, Fisher Scientific, Fair Lawn, NJ) was the extracting solvent.

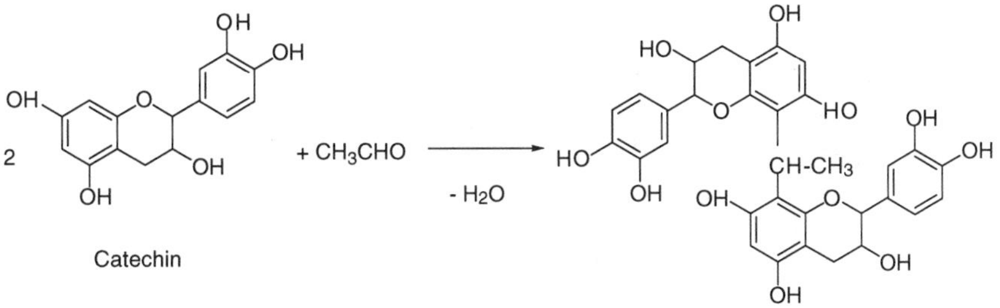

Figure 1. Polymerized product of acetaldehyde-phenol reactions (adapted from Ref. 26).

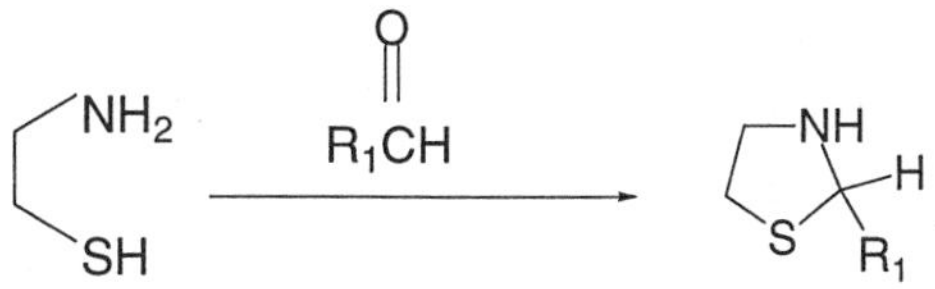

Figure 2. Chemical derivatization reaction of saturated aldehydes with cysteamine to form stable thiazolidine compounds.

Derivatization and Extraction. Modifications of the procedures of Ebeler et al. (49) were used for all aldehyde analyses. Briefly, 3.0 mL of wine were mixed with 60 μL of internal standard (10 mg 2,4,5-trimethylthiazole/mL in 10% aqueous ethanol) and 1 mL of 0.03 M aqueous cysteamine (pH 8.5); the pH was adjusted with HCl or NaOH (pH's from 2-10 were evaluated as discussed below). Following reaction at room temperature for 1 hour, the pH was re-adjusted to 8.5 and the solution was extracted two times with 1.5 mL of chloroform; the chloroform layer was removed each time and then combined to give a total of 3.0 mL of extract. Samples were injected onto a gas chromatograph fitted with either a mass spectrometer or nitrogen phosphorous detector. Peak area ratios of the internal standard to the analyte were used for all quantitative calculations.

Occassionally, emulsions form during extraction; centrifuging is normally sufficient to break the emulsion. In addition, a drying agent (anhydrous sodium sulfate) is often added after extraction since transfer of chloroform without water can be difficult.

Gas Chromatographic Conditions. All analyses were performed on a Hewlett Packard 5890 GC equipped with a 5970 Mass Selective Detector or a Hewlett Packard 6890 GC equipped with a Nitrogen Phosphorous Detector (Hewlett Packard, Inc., Avondale, PA). A DB 35 (35% phenyldimethylpolysiloxane), 30 m x 0.25 mm ID x 0.25 μm column (J & W Scientific, Inc., Folsom, CA) was used for all analyses. Carrier gas was helium at a linear velocity of 30 cm/sec. Samples were analyzed using split injections (split ratio = 30:1) with injector and detector (NPD) temperatures of 260°C and 250°C, respectively. Oven temperature programming was as follows: initial temperature of 80°C for 1 min; increase temperature at 3.5°C/min to 115°C; increase at 15°C/min to 180°C; increase at 60°C/min to 190°C; hold at 190°C for 6 min.

Standard Curve. Wines or model solutions (10% aqueous ethanol) were spiked with a mixture of eleven aldehyde standards for purposes of optimizing and evaluating the method. The aldehydes used included the C1 through C9 saturated, straight chain aldehydes and two branched chain aldehydes, 2-methyl-1-propanal and 3-methyl-1-butanal. Aldehydes were spiked to give concentrations between 0.1 and 30 μg/mL.

A standard curve was prepared for each aldehyde for all quantitative analyses. Known amounts of each aldehyde were added to 10% ethanol to give concentrations of 1, 4, 8, 12, and 16 μg/mL. Peak area ratios of aldehyde to IS were used to construct a linear standard curve for each aldehyde.

Results and Discussion

Comparison of Mass Spectrometer (MS) and Nitrogen Phosphorous Detectors (NPD). Using the MS detector in full scan mode, a limit of detection of 10 μg/mL was obtained. When reacted with cysteamine, each aldehyde forms a thiazolidine derivative which gives characteristic ions at m/z 56 and 88 in the mass spectrum. Using these ions as well as the molecular ion (or another characteristic fragment ion when a molecular ion was not present; Table 2) in the selected ion monitoring mode decreased the limit of quantitation to ~2 μg/mL. The limit of detection was also in this range (1-2 μg/mL). The MS detector was useful for confirming peak identities.

The NPD allowed an overall increase in sensitivity by a factor of ~10. Nonanal gave a limit of detection of 0.14 μg/mL with a signal to noise ratio of ~3:1. However, the detector response to the thiazolidine compounds was not linear below 1 μg/mL, the limit of quantitation for this detector. Therefore, both the MS and the NPD gave similar limits of quantitation; however, because the limit of detection was lower for the NPD, this detector was used for all subsequent analyses.

Overall limits of detection and quantitation may be decreased by concentrating the 3.0 mL extract under Nitrogen. We have not evaluated this possibility at this time.

Recoveries of Aldehydes Spiked in Wines. Aldehydes were spiked into wines at a concentration of 2×10^{-4} M (6 - 28.4 µg/mL), derivatized, and extracted as described above. The base wines included a Chardonnay, a Symphony, and a Cabernet Sauvignon. Aldehyde levels in unspiked wines were also determined, and all recoveries were corrected for the amount of aldehyde initially present in the wines. Average recovery for all aldehydes was 112.8% with an overall coefficient of variation of 16% (Table 3). Variable recoveries for octanal and nonanal may be due to limited solubility of these compounds in the matrix.

Effect of Derivatization pH on Aldehyde Recovery. Binding of aldehydes to other wine components (SO_2, phenols, etc.) is highly pH dependent, therefore the effect of pH on derivatization efficiency was evaluated. Following addition of aqueous cysteamine to spiked wine samples, the pH was adjusted to 2, 8, or 10, and the solutions were allowed to react for 1 hour. The pH of all samples was then re-adjusted to 8.5, and the samples were extracted and analyzed as described above. Initial results indicated that no consistent differences in recovery at the different pH's were observed, however, overall variability appeared greater at pH 2. These results provide preliminary evidence that the total aldehyde concentration (free plus bound) is measured with this procedure. Further studies with model solutions containing added SO_2 and phenols are needed to fully evaluate this result.

We did not evaluate the effect of both derivatization and extraction at different pH's. Although bound aldehydes may not be hydrolyzed and derivatized at acidic pH's, re-adjusting the pH to >8.5 just before extraction, as described above, may result in rapid hydrolysis and derivatization during the extraction process. By completing both derivatization and extraction at lower pH's (pH 6-7) it may be possible to estimate the amount of free (unbound) aldehydes. However, analysis at a pH lower than this is probably not feasible as Hayashi et al. (47) and Yasuhara and Shibamoto (45) observed a significant decrease in overall derivatization efficiency at pH's less than 6.

Aldehyde Levels in Different Varieties and Styles of Wine. The derivatization procedure described above was used to determine aldehyde levels in several different wines (Table 4). The wines were made in the UCD Department of Viticulture and Enology winery using standard procedures. As expected, acetaldehyde was the predominant aldehyde in all samples, with highest levels observed in the Sherry (Table 3). The acetaldehyde concentrations are consistent with those obtained by enzymatic analysis of acetaldehyde in table wines and Sherries (38).

Small amounts of formaldehyde were observed in all of the wines, again with the highest levels observed in Sherry (Table 4). Reported formaldehyde concentrations should be considered to be approximations however, as the exact formaldehyde concentration of the standard (~37%) was not determined for this study.

Trace amounts (<1 µg/mL) of many of the higher molecular weight aldehydes (C7 - C9) were also observed in all of the wines except the Cabernet Sauvignon. These results are consistent with those of Sponholz (50) who observed <1 µg/mL of propanal, isopropanal, propenal, butanal, isobutanal, pentanal, butenal, and hexanal in German Riesling wines using a Dinitrophenylhydrazine (DNPH) derivative and gas chromatographic analysis.

Effects of SO_2 Addition during Fermentation. Separate 4.5 gallon carboys containing Chardonnay juice were inoculated with *Saccharomyces bayanus*, followed one hour later by treatment with 0 ppm SO_2 (control), 50 ppm SO_2, or 200 ppm SO_2. Each treatment was done in duplicate for a total of six separate carboys. The wines

Table 2. Retention times and characteristic mass spectral ions for aldehydes derivatized with cysteamine to form thiazolidine derivatives.

Aldehyde	Thiazolidine Derivative	Retention time (min)	Characteristic ions (*m/z*)
Formaldehyde	Thiazolidine	5.37	56, 88, 59
Acetaldehyde	2-Methylthiazolidine	5.71	56, 88, 103
Propanal	2-Ethylthiazolidine	8.22	56, 88, 117
2-Methyl-1-propanal	2-Isopropylthiazolidine	9.97	56, 88, 131
Butanal	2-Propylthiazolidine	11.3	56, 88, 131
3-Methyl-1-butanal	2-Isobutylthiazolidine	12.7	56, 88, 145
Pentanal	2-Butylthiazolidine	13.8	56, 88, 145
Hexanal	2-Pentylthiazolidine	15.52	56, 88, 159
Heptanal	2-Hexylthiazolidine	16.91	56, 88, 126
Octanal	2-Heptylthiazolidine	18.44	56, 88, 140
Nonanal	2-Octylthiazolidine	20.32	56, 88, 159
2,4,5-Trimethylthiazole		6.88	59, 127

Table 3. Recoveries of aldehydes spiked into table wines.

Aldehyde	Mean Recovery (%)	S. D.
Formaldehyde	98.3	4.5
Acetaldehyde	85.9	3.2
Propanal	112.5	5.0
2-Methyl-1-propanal	117.5	11.9
Butanal	118.8	6.3
3-Methyl-1-butanal	112.5	8.7
Pentanal	112.5	8.7
Hexanal	102.5	8.7
Heptanal	115.0	2.6
Octanal	151.2	42.5
Nonanal	113.8	25.3

Values represent a minimum of three replications. Aldehydes were spiked into wines at a concentration of 2×10^{-4} M (6 - 28.4 µg/mL), derivatized, and extracted as described in text. Aldehyde levels were corrected for the amount of aldehyde initially present in the unspiked wines.

Table 4. Mean aldehyde levels determined in three different table wines and a Sherry*.

Aldehyde	1994 Chardonnay (μg/mL)	1993 Symphony (μg/mL)	1993 Cabernet Sauvignon (μg/mL)	Sherry (20 year) (μg/mL)
Formaldehyde	< 1	< 1	< 1	1.4 ± 0.09
Acetaldehyde	4.1 ± 2.3	6.2	5.8 ± 2.6	120.3 ± 2.5
Propanal	--	--	--	--
2-Methyl-1-propanal	--	--	--	5.3 ± 0.1
Butanal	--	--	--	--
3-Methyl-1-butanal	--	--	--	< 1
Pentanal	--	--	--	--
Hexanal	--	--	--	--
Heptanal	< 1	< 1	--	< 1
Octanal	< 1	< 1	--	< 1
Nonanal	--	< 1	--	--

*All values represent the average of two or more analyses.
< 1 indicates that aldehyde levels below the limit of quantitation (1 μg/mL) were detected.
-- indicates that aldehyde levels were below the limit of detection (0.1 μg/mL).

were allowed to ferment to dryness and then stored at 4°C for three weeks; samples were removed and analyzed for aldehydes as described previously.

As expected, aldehyde levels increased as the amount of SO_2 added increased (Table 5). Addition of SO_2 during fermentation has been shown to produce higher levels of acetaldehyde in the wines, possibly through inhibition of the aldehyde dehydrogenase enzyme (13). However, the longer chain saturated aldehydes, C6 - C9, appeared to decrease with the addition of SO_2 (Table 5). These results are consistent with those reported by Herraiz et al. (19) who observed that longer chain aldehydes are not as readily reduced and excreted by yeast during fermentation. Joslyn and Ough (9) also observed that addition of SO_2 decreased the amount of C6 and greater aldehydes formed by oxido-reductase enzymes.

Effect of SO_2 Addition Prior to Bottling and Storage of Chardonnay. Chardonnay grapes were crushed and pressed without SO_2, inoculated with *Saccharomyces bayanus*, and fermented to dryness at 50°C. A secondary malolactic fermentation was performed until disappearance of malic acid as determined by paper chromatography. After ~4 months storage on the yeast lees, the wine was racked and sterile filtered. Immediately prior to bottling the wine was divided into two lots: one lot was bottled after the addition of 30 ppm SO_2, the other was bottled without any added SO_2. After one year of storage at 56°F, formaldehyde, acetaldehyde, and 2-methyl-1-propanal were detected in the wines (Table 6). Interestingly, there were no differences in the amounts of these aldehydes between the two treatment groups, although an informal sensory investigation showed them to have quite different overall sensory characteristics and the wine which did not have added SO_2 had a slightly darker color than the wine with added SO_2. Further investigations of the effects of SO_2 on flavor changes in white wines during bottle storage are necessary to fully understand the observed results.

Conclusions and Further Work

The cysteamine derivatization procedure provided a sensitive method for quantitating volatile, saturated aldehydes (C1 - C9) in wine. Using NPD detection, a limit of quantitation of 1 µg/mL (Signal:Noise = 3:1) was obtained with average recoveries of 113% (Coefficient of Variation = 16%). With this method we were able to show differences in aldehyde levels in wines as a function of grape variety and processing conditions. Although acetaldehyde was observed in the highest concentrations, other aldehydes were also often present. The method now offers the opportunity to accurately evaluate the effects of fermentation and storage conditions on aldehyde concentrations in wines.

Further studies are planned to investigate the following parameters:

- Use of pure thiazolidine standards to prepare standard curves and evaluate derivatization and extraction efficiency. The exact purity of commercial aldehyde reagents (particularly formaldehyde) is difficult to determine. Thiazolidines can either be purchased commercially (with exactly known purity) or easily synthesized and purified and would provide improved precision in preparing the standard curve and calculating concentrations.

- Use of model solutions to determine the effects of SO_2 and phenolic composition on aldehyde recovery and precision.

- An evaluation of the effect of extraction at various pH's to determine the feasibility of using this procedure to measure both bound and free aldehydes.

- A comparison of results using this method and the standard AOAC distillation/titration procedure (AOAC Methods 967.10 and 972.09).

- Application of the derivatization and extraction procedure to juices and musts.

Table 5. Aldehyde levels in wines fermented with and without SO_2*.

Aldehyde	Control (μg/mL)	50 ppm SO_2 (μg/mL)	200 ppm SO_2 (μg/mL)
Acetaldehyde	71.5 ± 14.3	119.8 ± 13.9	256.0 ± 14.6
Propanal	--	--	< 1
2-Methyl-1-propanal	--	--	--
Butanal	--	--	--
3-Methyl-1-butanal	--	--	1.1 ± 0.02
Pentanal	--	--	--
Hexanal	< 1	--	--
Heptanal	< 1	--	--
Octanal	3.3 ± 0.4	2.5 ± 0.08	1.7 ± 0.6
Nonanal	< 1	--	--

*All values represent the average of two or more analyses.
< 1 indicates that aldehyde levels below the limit of quantitation (1 μg/mL) were detected.
-- indicates that aldehyde levels were below the limit of detection (0.1 μg/mL).

Table 6. Effect of SO_2 additions prior to bottling on aldehyde formation in stored (56°F) Chardonnay

Aldehyde	With SO_2 (μg/mL)	Without SO_2 (μg/mL)
Formaldehyde	< 1	< 1
Acetaldehyde	17.1 ± 0.07	16.9 ± 0.53
2-Methyl-1-propanal	< 1	--

All values represent the average of three analyses.
< 1 indicates that aldehyde levels below the limit of quantitation (1 μg/mL) were detected.
-- indicates that aldehyde levels were below the limit of detection (0.1 μg/mL).
Aldehydes not listed in the table were below the limit of detection (0.1 μg/mL).

178

- Further evaluation of the effects of fermentation and storage conditions on aldehyde concentration.

Acknowledgments

Financial support of the American Vineyard Foundation and the NIEHS Superfund Basic Science Research Program (#ES04699) is gratefully acknowledged. The authors would like to thank Courtney Siverson, Dr. Jim Lapsley, and Ernie Farinias for technical help and donation of wine samples.

Literature Cited

1. Zoecklein, B. W.; Fugelsang, K. C.; Gump, B. H.; Nury, F. S. *Wine Analysis and Production*; Chapman and Hall: New York, 1995, 621 pp.
2. Nagel, C. S.; Baranowski, E. S.; Baranowski, J. D. In *University of California, Davis Grape and Wine Centennial Symposium Proceedings*; University of California-Davis: Davis, CA, 1982; pp. 235.
3. Auerbach, C.; Moutschen-Dahmen, M; Moutschen, J. *Mutation Res., 1977, 39,* 317.
4. Marnett, L. J; Hurd, H. K.; Hollstein, M. C.; Levin, D. E.; Esterbauer, H.; Ames, B. N. *Mutation Res., 1985, 148,* 25.
5. Benedetti, A.; Comporti, M. *Bioelectrochemistry & Bioenergetics, 1987, 18,* 187.
6. Feinman, S. In *Formaldehyde Sensitivity and Toxicity*; Feinman, S. E., Ed; CRC Press: Boca Raton, FL, 1988; pp. 197-204.
7. *Merck Index, 15th Ed.,* Merck & Co., Inc.: Whitehouse Station, NJ, 1996.
8. Cayrel, A.; Crouzet, J.; Chan, H. W. S.; Price, K. R. *Am. J. Enol. Viticult., 1983, 34*(2), 77.
9. Joslin, W. S.; Ough, C. S. *Am. J. Enol. Viticult., 1978, 29*(1), 11.
10. Herraiz, T.; Herraiz, M.; Reglero, G.; Martin-Alvarez, P. J.; Cabezudo, M. D. *J. Agric. Food Chem., 1990, 38,* 969.
11. Baumes, R.; Bayonove, C; Barillère, J. M.; Escudier, J. L., Cordonnier, R. *Conn. Vigne Vin, 1988, 22*(3), 209.
12. Bitteur, S.; Tesniere, C.; Sarris, J.; Baumes, R.; Bayonove, C.; Flanzy, C. *Am. J. Enol. Viticult., 1992, 43*(1), 41.
13. Ough, C. S.; Amerine, M. A. *Methods for Analysis of Musts and Wines,* Wiley and Sons: New York, 1988, 377 pp.
14. Millán, M. C.; Moreno, J.; Medina, M.; Ortega, J. M. *Microbios, 1991, 65,* 87.
15. Di Stefano, R; Ciolfi, G. *Riv. Vitic. Enol., 1982, 35,* 474.
16. Nykänen, L. *Am. J. Enol. Vitic., 1986, 37*(1), 84.
17. Millán, C.; Ortega, J. M. *Am. J. Enol. Vitic., 1988, 39*(2), 107.
18. Amerine, M.A.; Berg, H. W.; Kunkee, R. E.; Ough, C. S.; Singleton, V. L.; Webb, A. D. *The Technology of Wine Making* 4th Edition. AVI Publishing Co., Inc.: Westport, CT, 1980.
19. Herraiz, T.; Tabera, J.; Reglero, G.; Cabezudo, M. D.; Martin-Alvarez, P. J. M.; Herraiz, M. *Belg. J. Food Chem. Biotechnol., 1989, 44*(3), 88.
20. Litchev, V. *Am. J. Enol. Vitic., 1989, 41*(1), 31.
21. Troton, D.; Charpentier, M.; Robillard, B.; Calvayrac, R.; Duteurtre, B. *Am. J. Enol. Vitic., 1989, 40*(3), 175.
22. Wildenradt, H. L.; Singleton, V. L. *Am. J. Enol. Vitic., 1974, 25*(2), 119.
23. Tenscher, A. C. *The Kinetics of Sulfite- Hydrogen Sulfite-Binding with Acetaldehyde and Pyruvic Acid.* M.S. Thesis, University of California, Davis, CA, 1986, 115 pp.
24. Hobley, T. J.; Pamment, N. B. *Biotechnology Techniques, 1997, 11*(1), 39.
25. Kaneda, H; Takashio, M.; Osawa, T; Kawakishi, S.; Koshino, S.; Tamaki, T. *J. Food Sci., 1996, 61*(1), 105.
26. Saucier, C.; Bourgeois, G.; Vitry, C.; Roux, D.; Glories, Y. *J. Agric. Food Chem., 1997, 45,* 1045.

27. Arnold, R. A.; Noble, A. C.; Singleton, V. L.*J. Agric. Food Chem.,* **1980**, *28,* 675.
28. Robichaud, J. L.; Noble, A. C. *J. Sci. Food Agric.,* **1990**, *53,* 343.
29. Ebeler, S. E. In *Recent Advances in Phytochemistry, Functionality of Food Phytochemicals,* Romeo, J. T., Ed., Plenum Press: New York, 1997; Volume 31.; In Press.
30. Williams, P. J.; Strauss, C. R. *J. Sci. Food Agric.,* **1975**, *26,* 1127.
31. Tuma, D. J.; Hoffman, T.; Sorrell, M. F. In *Advances in Biomedical Alcohol Research,* Kalant, H; Khanna, J. M.; Israel, Y, Eds., Pergamon Press, New York, 1991, pp. 271-276.
32. Nichols, R., de Jersey, J.; Worrall, S.; Wilce, P. *Int. J. Biochem.,* **1992**, *24*(12), 1899.
33. Schmidt, R. H.; Davidson, S. M.; Bates, R. P. *J. Food Sci.,* **1983**, *48,* 1556.
34. AOAC, *Official Methods of Analysis;* Association of Official Analytical Chemists, Inc.: Arlington, VA, 1996, 16th Edition.
35. Owades, J. L; Dono, J. M. *J. Assoc. Offic. Anal. Chem.,* **1968**, *51*(1), 148.
36. Ballesta, P. L.; Olea, M. R.; García-Villanovo, R. *Anal. Bromatol.,* **1980**, *32*(4), 367.
37. Esterbauer, H.; Zollner, H. *Free Rad. Bio. Med.,* **1989**, *7,* 197.
38. McCloskey, L. P.; Mahaney, P. *Am. J. Enol. Vitic.,* **1981**, *32*(2), 159.
39. Priefert, H.; Steinbüchel, A. *Biotechnology Letters,* **1993**, *15*(5), 443.
40. Noguer, T.; Marty, J. L. *Enz. Microb. Technol.,* **1995**, *17,* 453.
41. Morita, H.; Inoue, H.; Tanabe, O. *J. Ferment. Technol.,* **1969**, *47*(5), 303.
42. de Revel, G; Bertrand, A. *J. Sci. Food Agric.,* **1993**, *61,* 267.
43. de Revel, G.; Bertrand, A. In *Trends in Flavour Research,* Maarse, H, van der Heij, D. G., Eds., Elsevier Science B.V.: Amsterdam, 1994, pp. 353-361.
44. Shibamoto, T. In *Flavors and Off-Flavors.* Charalambous, G., Ed., Elsevier Science B.V.: Amsterdam, 1989, pp. 471-483.
45. Yasuhara, A.; Shibamoto, T. *J. Chromatogr.,* **1991**, *547,* 291.
46. Yasuhara, A.; Shibamoto, T. *J. Assoc. Off. Anal. Chem.,* **1989a**, *72*(6), 899.
47. Hayashi, T.; Reece, C. A.; Shibamoto, T. *J. Assoc. Off. Anal. Chem.,* **1986**, *69*(1), 101.
48. Yasuhara, A.; Shibamoto, T. *J. Food Sci.,* **1989b**, *54*(6), 1471.
49. Ebeler, S. E.; Hinrichs, S. H.; Clifford, A. J.; Shibamoto, T. *J. Chromatogr. B.,* **1994**, *654,* 9.
50. Sponholz, W.-R. *Z. Lebensm. Unters. Forsch,* **1982**, *174,* 458.
51. Fenaroli, G. *Fenaroli's Handbook of Flavor Ingredients,* 2nd Ed., Furia, T. E.; Bellanca, N., Eds.; CRC Press: Cleveland, OH, 1975, Vol. 2.

Chapter 14

Volatile and Odoriferous Compounds in Barrel-Aged Wines: Impact of Cooperage Techniques and Aging Conditions

Pascal Chatonnet

Seguin-Moreau Cooperage, Faculty of Enology, University Victor Segalen, Bordeaux II, 351, Cours de la Libération, 33405 Talence cedex, France

Oak (*Quercus sp.*) has long been used for aging fine wines and brandies thanks to its physical and chemical properties. Barrel aging is a major factor in enhancing and stabilizing the wines. However, only certain species of oaks and only those from some geographical regions, have proved to be really interesting. Cooperage operations, especially drying methods and barrel toasting techniques, can considerably change the wood's composition in term of extractable components, especially volatile and odoriferous compounds. In addition to the oak's intrinsic characteristics and the barrelmaking process, the way in which the barrel is used to aged the wines can also greatly influence the quality and taste of wine.

Great wines often spend several months, or even years, in oak barrels before being bottled. This type of container has been used over the centuries and remains in widespread use because it enhances the intrinsic qualities of many wines. The practice of using wood for transporting and aging wines is both ancient and recent. The invention of the cask or barrel has been attributed to Celtic genius in Northern and Eastern Europe. In spite of the fact that the first wines were made in earthen jars (*dolia*), containers made from goatskin were the first to be used to transport liquids, especially wine. As the wine and oil trades developed around the Mediterranean, thanks to the Phoenicians and then the Greeks, Canaanitic amphorae made from baked earth gradually replaced goatskin. The first traces of the use of wooden casks date from the end of the 5th century BC, at the height of the Etruscan civilization. These casks transported wines from the Po valley to Rome. During the 2nd century AD, amphora production decreased and wooden barrels gradually began to take their place. By the end of the Antonine era, however, barrels were clearly preferred. For over 1,700 years, the containers most frequently used for wine production, transport and storage have been made of wood. The Gauls, renowned for their expertise in carpentry, certainly facilitated the development barrel-making, but it is unsure whether or not they actually invented it.

Over the years, many tree species have been used to make barrels, but oak (genus *Quercus*) quickly proved to be the best suited to aging fine wines in general and Bordeaux in particular. Among the various oak species, only European sessile oak (*Q. sessilis*), split along the grain rather than swan, is well-adapted to making the staves of barrels for aging great wines. White American oak (*Q. alba*) may also be used in certain circumstances. French oak only started making a name for itself at the beginning of this century. Up until then, the quality and quantity of wood available in France were insufficient. French coopers long had to use imported wood from Northern and Eastern Europe, and even North America, because of widespread deforestation and the need to use old trees (150 to 200 years old). Local wood (often pedunculate oak) was set aside for lower quality wines. Since then, French tree farmers have worked long and hard to develop the greatest oak forest in Europe, both in terms of quantity and quality (14 millions ha, 8.5 millions ha of oaks, 140,000 ha of haute-futaies). Their efforts have aided vine growers around the world.

In 1858, the Bordeaux region established a precise definition of the 'bordelaise' barrel. Its particular shape and appearance have since become common for aging great red and white wines all over the world. The 'bordelaise', with a volume of 225 litres, was originally made with chestnut hoops, which are now metal. These barrels have a distinctive curved shape that makes them easy to maneuver, either empty or full. This was a considerable advantage at a time when wine was transported by boat.

Winemakers' attitudes towards the use of wood have developed considerably over the years. In the past, barrels were seen simply as containers, preferably as neutral as possible (a "taste of the barrel" was thought to be a defect). Barrels were abandoned in Europe to a great extent in favor of cement or stainless steel vats in the mid-20th century. Wood has since returned to cellars all over the world. However, barrels are seen in a new light. We now know the precise effect of oak on winemaking, and barrels are no longer considered as simple containers. Oak is not an inert material. Most of the changes that occur in barrel-aged wine are indispensable for further aging in bottle. Today this seems obvious, but it was not so in the past.

For a long time, barrels were the only containers used to transport wines. This came to an end when the market called for wines bottled in the region of production, later became required by law. The whole conception of aging wine changed. In fact, thanks to progress in winemaking and recent research into the chemical composition of oak and its interaction with wine, modern enologists and cellarmasters can draw even further benefit from aging wine in barrel. Winemakers are much more aware of the impact of cooperage techniques on wine flavor and quality. The enological significance of the selection of oak types/origins, wood-drying and seasoning, as well as barrel toasting are now better understood. Coopers and winemakers are able, therefore, to cooperate more efficiently in pursuit of perfection.

1- Composition of oak wood and botanical origins

We have today a better understanding of the composition of oak of various origins. Oak is a naturally aromatic wood. This is largely due to the presence of a highly aromatic lactone, ß-methyl–γ–octalactone [1], which is specific to the *Quercus* genus and responsible for oak's typical aroma (1). Beside volatile and aromatic substances, there are also stable polyphenolic compounds named ellagitannins (2,3). There are essentially two types of oak tannins: vescalagin and castalagin, both polyesters of hexahydroxydiphenic acid and glucose.

1

European sessile oak, generally from fully-grown trees, has a loosely-knit structure, with a slow, even growth rate. It has relatively little tannin for wood with such high aromatic potential. Pedunculate oak, on the other hand, has a much greater and more irregular growth rate. It is more compact, rich in tannin and lower in aromatic substances. White American oak (Q. alba) is quite compact, and is sawn rather than split. In contrast, European species of oak must always be split lengthwise along the grain of the wood in order to prevent sap leaking onto the staves. Tylosis membranes which block the vessels of the heartwood, are different in American oak (4). They are efficient in preventing liquid from flowing through sawed vessels, whereas these same membranes in European species are relatively permeable. As a result, the usable yield from American oak is approximately twice (50 % minimum vs 25 % maximum) that of European oak. In addition, American white oak is lacking in tannic compounds, but rich in methyl-octalactone. It has an aromatic potential which is, on average, equal to or twice that of sessile oak (5).

Beside positive aromas, oak wood can give in some circumstances disagreeable off-flavors as « sawdust » aroma. A combination of gas chromatography, mass spectrometry and olfactive detection were used to isolate several aromatic zones possessing odors reminiscent of the various nuances of the "sawdust" or « plank » aromas found in certain wines aged in new oak barrels (6). (E)-2-nonenal (2) is the molecule largely responsible for this disagreeable odor (figure 1).

2

Its concentration varied considerably from one wood sample to another. In addition, 3-octen-1-one was present in certain untreated wood samples and may reinforce the unpleasant odor resulting from some others unsaturated aldehydes. By measuring (E)-2-nonenal, after derivation by O-(2,3,4,5,6-pentaflurobenzyl)-hydroxylamin (PFBOA), in wines more or less affected by the "sawdust" aroma, it was possible to observe a satisfactory correlation between the intensity of this off-flavor and the concentration of this compound in wine. (E)-2-nonenal had a fairly low perception threshold (180-200 ng/l for 50 % of the tasters) and the presence of a concentration about three times higher than this (approximately 600 ng/l) alter dramatically the quality of a red wine's aroma. Concentrations of over 2 µg/l were perceived as an important off-odor in the majority of cases, although, depending on

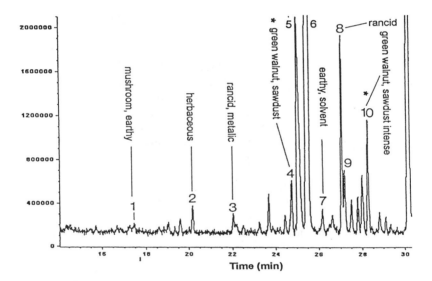

Figure 1. Identification of volatile compounds responsible for the « sawdust » aroma of oak wood by coupling gas chromatography and olfactive detection

the wine's aromatic intensity, it could be considered slight or very marked. We measured up to 9.5 µg/l in certain samples.

Volatile aldehydes, and (E)-2-nonenal in particular, had already been identified as the cause of "rancid odors" in beer (7,8). These substances result from the oxidation of unsaturated fatty acids. The direct precursor of (E)-2-nonenal and others carbonyl components is linoleic acid ($C_{18:2}$ Δ 9,12) (9). Volatile aldehydes may be derived from fatty acids in various manners. Chemical auto-oxidative factors would seem to provide the most likely explanation for the presence of these components in oak stave wood after seasoning in the open air. On the other hand, enzymatic factors may explain the presence of these components while the tree is still standing or immediately after it has been cut. Additional research is necessary to pinpoint the exact formation and accumulation mechanisms of these molecules in the wood.

2- Influence of the geographical origins on the composition of European oaks

In France, coopers use oak from several French forests located in four main regions (figure 2). According to experienced coopers, each geographical area produces wood with specific characteristics, capable of contributing distinctive flavors and aromas to wines and brandies. We measured differences in compounds extracted in a dilute alcohol medium from oak wood of known origins following natural, outdoor seasoning in order to determine whether these reputed differences had a basis in fact.

Table I shows the major physicochemical characteristics of wood from the four main French regions. It highlights the, in some cases striking, differences in the composition of oak wood according to its geographical origin (10). Wood in the Limousin group stood out considerably from the other types, due to their concentration of extractable, non-volatile substances, especially phenolic compounds. From this point of view, there is relatively little difference between the other regions. Among the four regions considered, oak from the Center had the lowest content of coloring matter.

The group from the Center had a distinctly higher methyl-octalactone and eugenol content than the other regions. The group from the Vosges, in the far north-east, was easily distinguishable, with a relatively high extractable methyl-octalactone concentration in relation to its eugenol content. The Burgundy group was more difficult to isolate, as it had no distinctive characteristics. It was fairly similar to the Center and Vosges groups in terms of its polyphenol content, whereas its low volatile compound content was more like that of the Limousin group.

Figure 3 shows the overall aromatic potential of oak from the various origins. This cumulative histogram presentation is intended simply as an illustration, as there is no real point in adding the aromatic indices together. Wood from the Center and Vosges groups seemed to be the most aromatic, whereas the Burgundy and Limousin groups were fairly similar and over three times less aromatic. According to our method of calculation, the methyl-octalactone content had the greatest impact on the overall aromatic potential.

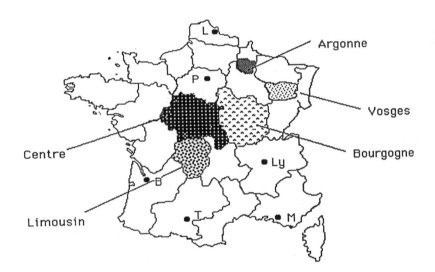

Figure 2. Location of the main regions in France supplying oak wood for the cooperage

Table I. Influence of the geographical origin on the composition of wood from French oaks seasoned naturally in the open air (7 samples per origin)

	Geographical origin			
Analytical parameter	Limousin	Center	Burgundy	Vosges
Total extractives (mg/g)	140 (7.2)	90 (15)	78.5 (1.7)	75 (3.9)
Total polyphenols (A_{280})	30.4 (1.8)	22.4 (2.9)	21.9 (2.8)	21.5 (1.7)
Coloring (A_{420})	0.040 (0.008)	0.024 (0.001)	0.031 (0.002)	0.040 (0.004)
Catechic tannins (mg/g)	0.59 (0.08)	0.30 (0.03)	0.58 (0.12)	0.30 (0.02)
Ellagic tannins (mg/g)	15.5 (1.5)	7.8 (1.4)	11.4 (2.5)	10.3 (0.8)
Methyl-octalactones (μg/g)	17 (15)	77 (24)	10.5 (4.5)	65.5 (12)
Eugenol (μg/g)	2 (1.40)	10 (4.50)	1.8 (0.80)	0.6 (0.020)

() :standard deviation

Sessile and pedunculate oaks grow throughout Europe, except for the far north. We focused on oak from southern Russia, at the foot of the Caucasus mountains, widely used in France at one time. Following analysis of several (naturally-seasoned) wood samples from various regions located north-west of the Caucasus (Adygea), we demonstrated the existence of wood with characteristics similar to those of the French standard woods (11). In this way, by comparing analytical profiles of Russian oaks with those of the French standard woods (figure 4), it was possible to identify certain forest reserves that would, *a priori,* be suitable for manufacturing wine barrels as they are similar to oaks from the Center region, whereas others are better suited for aging brandies as they resemble Limousin oak. Subject to rigorous selection, carefully-controlled drying and seasoning, and perfect manufacture, these woods could, in the very short term, represent a very useful source of supply for cooperage. A series of tests on several wine estates in different appellations, both in France and Australia, has confirmed the suitability of these woods, both for aging red wines as well as fermenting whites (12).

3- Influence of the seasoning and the aging of the oak wood

In order to be made into barrels, oak must first be dried. This operation is traditionally done by stacking wood and storing it outside for several years. Our research has shown that, more than merely drying the wood out, the seasoning of oak also has an important effect on barrel quality, and thus on wine. During this time, due to changes in humidity levels and contact with oxygen in the air, there is a constant decrease in the wood's oligomeric ellagitannins, which can be responsible for unpleasant, bitter flavors (13). Oak trees, when felled, often have very little aroma. However, there is a considerable increase in methyl-octalactone content when the wood is exposed to the elements in order to age (figure 5). Methyl-octalactone molecules exist in two isomeric forms. The (3S,4S) form is four times as aromatic as the (3S,4R) form (10, 14, 15). During natural wood-drying, there is a net increase in (3S,4S)-methyl-octalactone due to the breakdown of their precursors (lipidic esters) (16). The *trans* form of the precursor is more stable than the *cis*. It results that the *cis* precursor is largely hydrolyzed to give free and aromatic (3S,4S)-methyl-octalactone during the natural seasoning of the staves (figure 6). The exact mechanisms at the origin of methyl-octalactones' precursors in wood and their hydrolysis are still unknown.

At the same time, there is also an oxidative breakdown of the lignin terminal molecular chain remaining after acidolysis simply in presence of water and organic acids. This results in the release of small quantities of phenolic aldehydes and volatile phenols (figure 7). Among those molecules that have been identified, vanillin and eugenol are the most odoriferous. But, in comparison with the quantities of these molecules which can be formed during the toasting of the barrels, contribution of seasoning is low (20 to 30%). Artificial drying of staves is much quicker. However,

Total polyphenols (A ₂₈₀)/ Aromatic potential (AP)

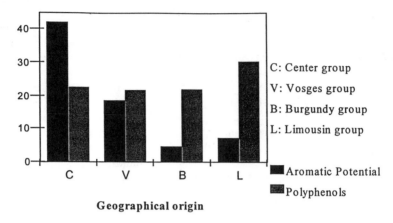

C: Center group

V: Vosges group

B: Burgundy group

L: Limousin group

■ Aromatic Potential

▨ Polyphenols

Geographical origin

Figure 3. Variations in aromatic potential (concentration of volatile compounds/perception thresholds) and polyphenolic content of oak wood according to geographical origin

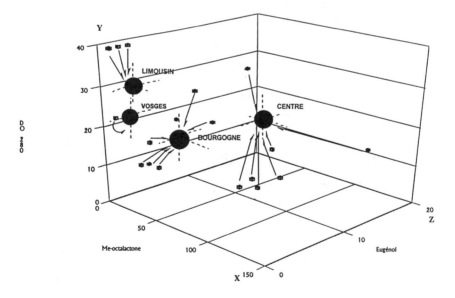

Figure 4. Comparison of the analytical profiles of Russian oaks (autonomous republic of Adygea, northwestern Caucasus, Russia) with the average profiles of the principal French reference samples by principal component analysis (5)

the rapid elimination of water and the considerable shortening of the storage period prevent the favorable development of the oak's chemical composition. In comparison with natural seasoning, artificially-dried oak can produce unpleasant flavors and aromas (piney/resiny) and possess a lower aromatic potential (figure 8), and is not suitable for aging fine wines.

The microflora present on the wood during its open air drying and aging has been studied (17, 18). A lot of fungi, some yeasts and bacteria are present on the wood and in the first millimeters of the staves (0-4 mm) in all the situations. The penetration of the fungi in the deepest parts of the wood needs time and water to allow a significant colonization. Staves seasoned less than three to five years, and not watered regularly, are only poorly colonized and never in the inner parts of the oak wood. The action of a large part of the microflora on the degradation of the ellagic tannins at the surface of the oak wood is demonstrated and positive for its quality ; its action more deeply is low and not efficient. The majority of the fungi identified are able to degrade the volatile and aromatic compounds present in the wood. So, the great increase in the quality of oak wood during its natural seasoning doesn't seems to be largely dependent of the microflora development. Physical and chemical transformations (lixivation, hydrolysis, oxidation) appear to be the most important phenomena responsible for the oak wood maturation

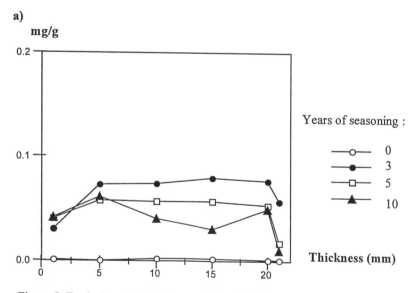

Figure 5. Evolution of (3S,4R) *trans* (a) and (3S,4S) *cis* β-methyl-γ-octalactone during oak seasoning

Continued on next page.

b)

mg/g

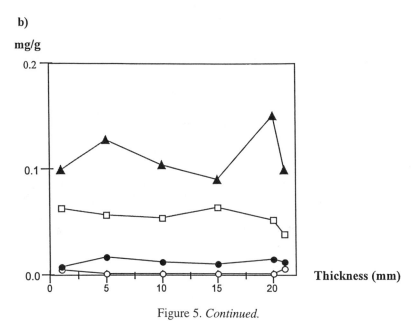

Figure 5. *Continued.*

a)

mg/g

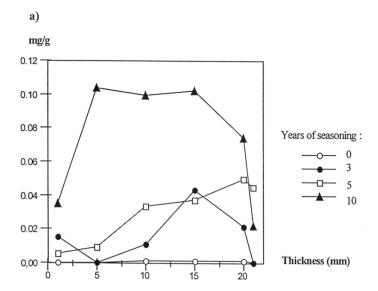

Figure 6. Evolution of *trans* (a) and *cis* (b) methyl-octalactone precursor during oak seasoning

Continued on next page.

b)

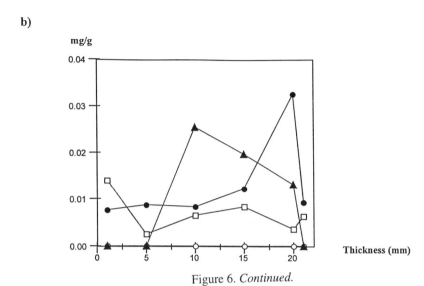

Figure 6. *Continued.*

a)

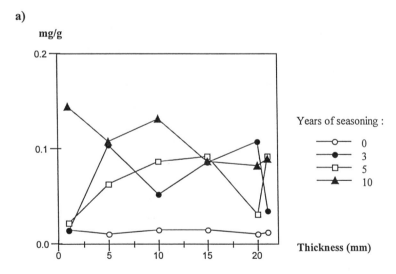

Figure 7. Evolution of vanillin (a) and eugenol (b) content of oak wood during seasoning

Continued on next page.

b)   mg/g

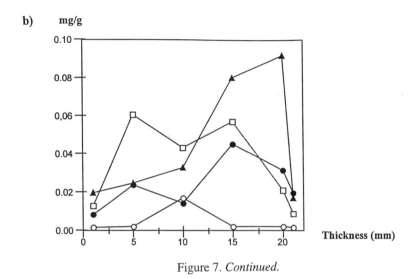

Figure 7. *Continued.*

Aromatic potential (olfactive units)

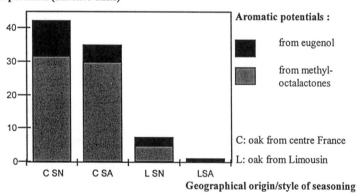

Figure 8. Aromatic potential (concentration of extractives/perception thresholds = olfactive units) of oak wood after natural seasoning during three years (SN) or artificial drying (3 months) (SA), adapted from (5)

4- Influence of heating on the wood aromas

After drying and aging, the planks of wood may be cut into staves and made into barrels. Toasting is a key stage in barrel-making. In fact, when wood is toasted for reasons beyond simply bending the staves, increasing the wood's temperature causes a significant change in the chemical composition of the barrel's inner surface. The heat helps to break down carbohydrate polymers (hemicelluloses) and phenolic polymers (lignins and ellagitannins), creating some molecules and eliminating others.

Lignin degradation produces several phenolic aldehydes in large quantities. Among these, only vanillin has a strong aroma, exactly like that of vanilla. Thermal degradation of the phenolic aldehydes produces a large number of volatile phenols. These compounds have "smoky" aromas, like guaiacol or 4-methyl-guaiacol, or "spicy" aromas, like eugenol.

The methyl-octalactone isomers and ellagitannins in wood before toasting tend to decrease as toasting increases ; to limit the excessive impact of methyl-octalactones on aroma with the american white oak, the coopers have generally increase time and intensity of the toasting (5). Depending on the length of toasting, coopers produce barrels of varying aromatic intensity and with different aromas. The peak in the formation of aromatic molecules during toasting varies according to the nature of chemical groups and to the way of heating by the cooperage.

Compounds such as furanic aldehydes (figure 9) caused by carbohydrate degradation, which are responsible for faint "toasty" aromas and vanillin (figure 10) with aroma of « vanilla », tend to be formed at a medium toast level.

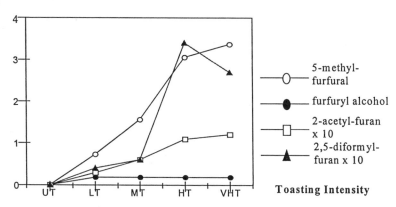

Figure 9. Influence of the toasting intensity on the furanic compounds of oak wood (UT : untoasted, LT : light toast, MT : medium toast, HT : high toast, VHT : very high toast)

mg/l

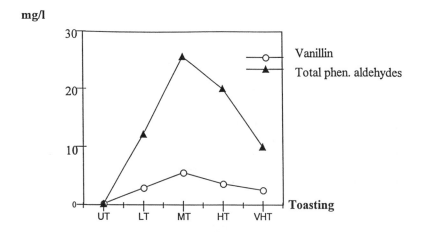

Figure 10. Evolution of phenolic aldehydes during the toasting of oak wood (fraction 0-5 mm of the inner face of the staves exposed to fire, 20 g/l ethanol 12 % v/v of wood)

Volatile phenols increase at a regular rate until a heavy toasting level is reached (figure 11). Under the action of the heat, the degradation of the precursor of méthyloctalactones can give free and odorous γ-lactones in the first millimeters of the staves. But if the intensity of toasting reach the high toast, there is a quick degradation and the quantity of extractables in wines decrease markedly.

μg/l

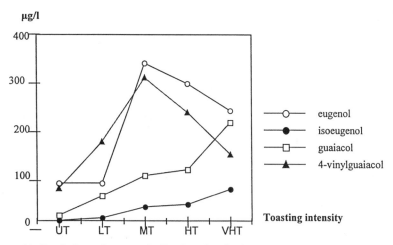

Figure 11. Evolution of some volatile phenols of oak wood with the toasting intensity

concentration in MOL equivalent (µg/g)

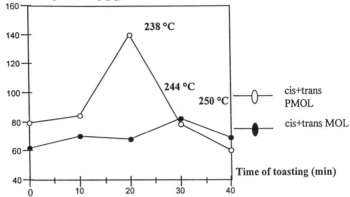

Figure 12. Evolution of methyl-octalactones (MOL) and its precursor (PMOL)during the heating of oak wood at the laboratory

The « medium toast », corresponding to a time of heating between 12 and 15 minutes after the bending, corresponds to the more complex and rich aromas (table II). As a result, the aroma and flavor of wine to be aged in barrel will be greatly influenced by wood origin and toasting intensity. Similarly, E-2-nonenal, the molecule responsible for the very unpleasant "sawdust" odor released by some types of oak, decreases with heavier toasting (figure 13).

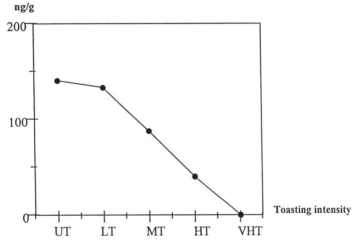

Figure 13. Evolution of E-2-nonenal during the toasting of oak wood

Identification of new molecules with "toasty" aromas

Toasted wood extracts and standard media heated in the laboratory were analyzed by gas chromatography and olfactory detection. Several "toasty" aromatic

Table II. Incidence of the origin of wood and intensity of toasting on the volatile compounds of a white wine aged nine months in new oak barrels

	Control	ALLIER				LIMOUSIN			
		L	M	H	VH	L	M	H	VH
Total polyphenols (A $_{280}$/PVPP)	3	4	3.9	3.9	3.8	5.2	4.3	4.7	4.4
Coloration (A $_{420}$)	0.1	0.12	0.13	0.13	0.08	0.42	0.47	0.48	0.48
(mg/l)									
furfural	0	0.9	3.6	4.9	3.5	1.8	2.55	4.8	4.3
5-methyl-furfural	0	0.8	1.1	0.75	0.5	0.9	0.95	0.8	0.4
furfuryl alcohol	0	0.5	5.1	4.8	4.2	4	3.6	4.3	1.8
Σ furanic compounds	0	2.2	9.8	10.45	8.2	6.7	7.1	9.9	6.5
(mg/l)									
trans methyl-octalactone	0	0.13	0.17	0.053	0.037	0.067	0.051	0.023	0.012
cis methyl-octalactone	0	0.29	0.14	0.089	0.114	0.095	0.095	0.055	0.058
Σ methyl-octalactones	0	0.42	0.31	0.142	0.151	0.162	0.146	0.078	0.07
(μg/l)									
guaiacol	2	10	18.5	38	65	6	12	21	33
4-methyl-guaiacol	0	10	14	24	29	10	11	14	18
4-vinyl-guaiacol	150	98	114	149	117	104	110	99	74
4-ethyl-4-guaiacol	0	9	9	14	15	4	4	4	13
eugenol	0	27	29	38	28	13	13	19	23
phenol + o-cresol	8	25	26	47	41	26	27	17	35
p-cresol	-	1	1	2	-	0	1	1	0
m-cresol	-	2	3	4	-	2	2	1	0
4-vinylphenol	300	197	206	319	210	187	211	214	131
(mg/l)									
vanillin	0	0.29	0.35	0.36	0.2	0.2	0.64	0.43	0.1
syringaldéhyde	0	0.49	0.69	1.4	1.8	0.27	0.4	-	-
Σ phenolic aldehydes	0	0.88	1.04	1.76	2	0.47	1.04	-	-

L: light toast, M: medium toast, H: high toast, VH: very high toast zones were pinpointed among the many chromatographic peaks. The combination of mass spectrometry and infra-red spectroscopy, then co-injection with pure reference products, either commercially available or synthesized in the laboratory, made it possible to identify the major compounds responsible for these aromas (19). By coupling gas chromatography and olfactory detection (figure 14) five reproducible "toasty" aromatic zones were identified (referred to as OZ).

Cyclotene (2-hydroxy-3-methyl-cyclopentenone, **3**) and maltol (3-hydroxy-2-methyl-pyranone, **4**) were easily identified as being responsible for OZ 1 and 3, with "burnt sugar" and "caramel" characteristics. These compounds have been previously identified in toasted oak wood by (20, 21).

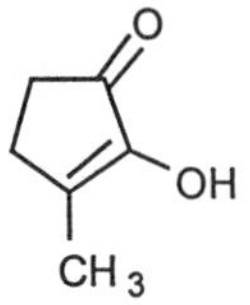

3 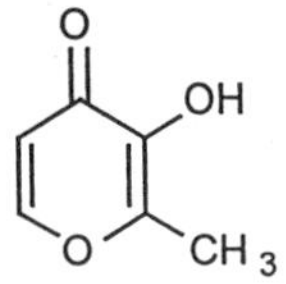   **4**

The compound responsible for OZ 2 has a very intense "sweet vanilla" odor, but it is co-eluted with guaiacol on Carbowax 20M, which prevents it from being easily detected. Pre-fractionation on silica gel, combined with mass spectrometry and infra-red spectroscopy, identified the molecule responsible: 2,3-dihydro-5-hydroxy-2-methyl-4(H)-pyran-4-one (**5**), or DHM.

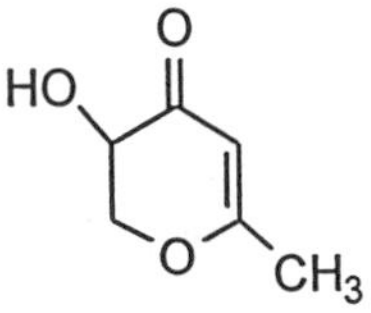

5

4-hydroxy-2,5-dimethyl-3(2H)-furan-3-one [6], furaneol or HDMF, has an intense, persistent "fruity-toasty" aroma, coinciding with OZ 4.

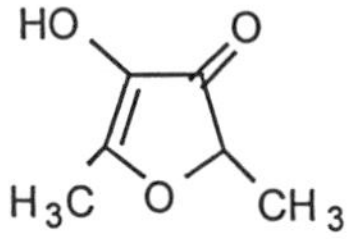

6

The aromatic zone ZO 5 has a "toasty" character with "fruity caramel" overtones, which we identified as 2,3-dihydro-3,5-dihydroxy-2-methyl-4(H)-pyranone (**7**), or DDMP.

198

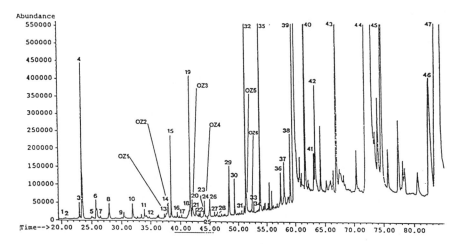

1: 2,5-dimethyl-pyrazine, 2: 2,6-dimethyl-pyrazine, 3: acetic acid, 4: furfural, 5: furanyl-1-ethanone, 6: benzaldehyde, 7: propionic acid, 8: 5-methyl-furfural, 9: butyrolactone, 10: hydroxy-benzaldéhyde, 11: 3,4-dimethyl-furanone-2(5H), 12: furanone-2(3H), 13: cycloten, 14: hexanoic acid, 15: guaiacol, 16: *trans* methyl-octalactone, 17: 2-phenyl-1-éthanol, 18: benzothiazol, 19: *cis* methyl-octalactone + 4-methyl-guaiacol, 20: maltol, 21: 2,5-diformyl--furan, 22: o-cresol, 23: phenol, 24: 4-ethyl-guaïacol, 25: 1H-pyrolle-carboxaldehyde, 26: octanoic acid, 27: p-cresol, 28: m-cresol, 29: eugenol, 30: isomaltol, 31: 4-vinylguaiacol, 32: syringol, 33: decanoic acid, 34: iso-eugenol, 35: 4-methyl-syringol, 36 dodecanoic acid, 37: 5-hydroxymethyl-furfural, 38: 4-allyl-syringol, 39: vanillin, 40: acetovanillon, 41: tetradecanoic acid, 42: propiovanillon, 43: butyrovanillon, 45: acétosyringon, 46: propiosyringon, 47: coniferaldehyde

Figure 14. Chromatogram (Carbowax 20M) of an extract of toasted oak wood
OZ represents the odorent zones with« toasty » aromas detected by olfactive detection

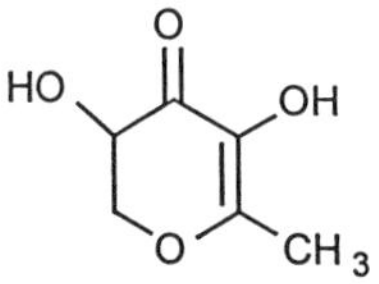

7

An additional aromatic zone OZ 6 is also present in some wood extracts and in models of Maillard reactions with proline and glucose. The molecule responsible for the interesting odor of « jam » and « burnt sugar » has been identified to 3,5-dihydroxy-2-méthyl-2,3-dihydro-4(H)-pyranone or hydroxymaltol [8].

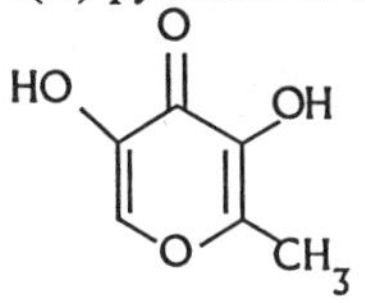

8

Compounds **5, 6, 7** and **8** were detected for the first time in toasted oak.

Origin of « toasty » compounds in the oak wood

None of the identified molecules are present in untoasted wood. Although these compounds are produced by the heat breakdown of simple or complex sugars in the wood, pyrolysis (direct heating) only creates some of the molecules identified. Much larger quantities of these molecules are formed when the sugars are heated in the presence of amino acid residue. As a result, the molecules that give a "toasty" aroma formed during the toasting of the wood are generated by a complex chain of reactions known as *Maillard reactions* (figure 15).

Cyclotene is formed from the less basic amadori intermediates and can be synthetized at a lower temperature than maltol (22). Maltol is the principal product of the thermal degradation of 1,4-disaccharides (23). DDMP may be a pivotal molecule in all these reactions. When it breaks down under heat, all the other reproducible "toasty" aromatic compounds that we have identified may be formed. DHM is formed from DDMP in acidic conditions because the water elimination is more easy with low pH (24). Furaneol can be formed by the thermal degradation of DDMP or by condensation of 1-hydroxy-2-propan-1-one, *via* acetylformoin. Hydroxymaltol can be easily transformed into maltol and DHM by dehydration. Acetylformoin, one of principal by-products identified in the thermal degradation of DDMP by (24) is not clearly identified in all the wood extracts. This molecule is not formed if the temperature is under 150°C or in presence of water. In the conditions of toasting at the cooperage, the temperature is largely over this limit (25) ; vapor is formed during the heating of wood and the coopers use small quantities of water for the bending of the staves and to prevent charring. So, it is probable that acetylformoin can be easily transformed in furaneol by dehydration in these particular conditions.

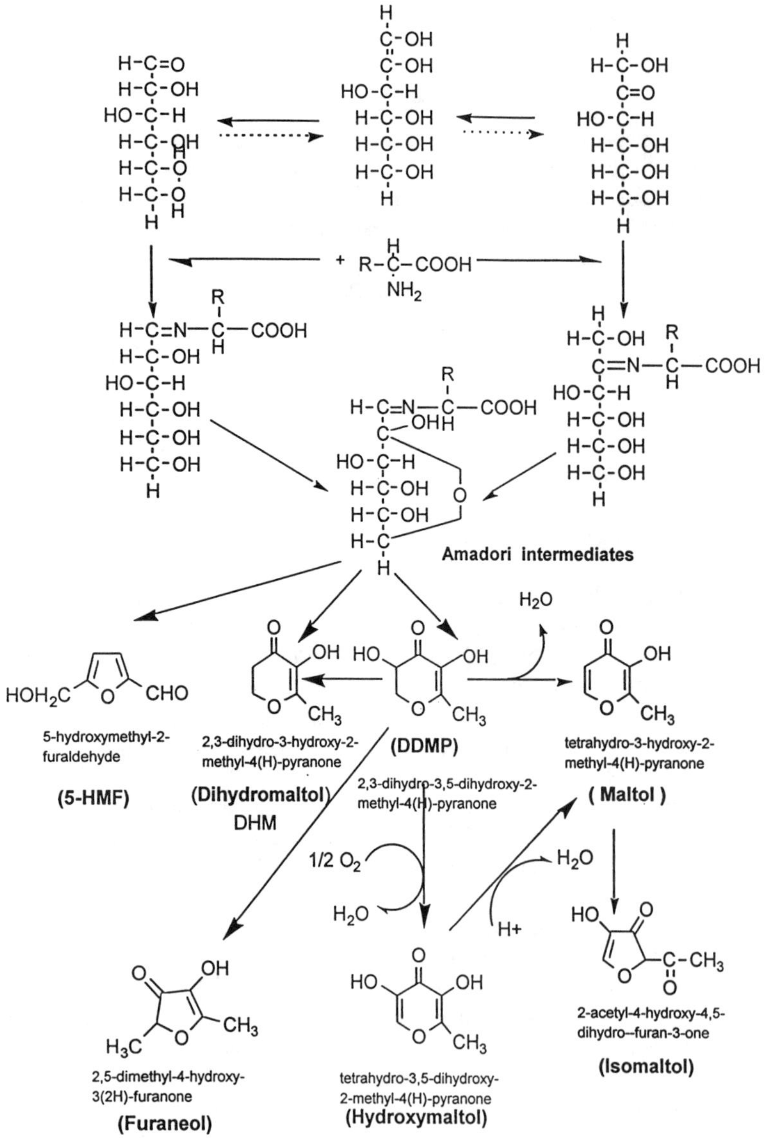

Figure 15. Formation of volatile compounds with "toasty" aromas by Maillard reactions between sugar molecules and amino acid residues when wood is heated

All of these compounds have similar aromas, reminiscent of « caramel » and « toast ». HODGE (26) considers that this similarity is due to the following enolic group in the molecule cycle :

$$\underset{H_3C}{\overset{R}{\diagdown}}C=\underset{\overset{|}{OH}}{C}-\underset{\overset{|}{R'}}{C}\overset{\diagup O}{}$$

The furanic aldehydes 5-(hydroxy-methyl)furfural and 2-furaldehyde, systematically present in the toasted wood, can be formed by the thermal degradation of 3-deoxyosone during sugar pyrolysis or Maillard reactions (27). They could also be formed from glyceraldehyde, coming from degradation of DDMP, by condensation with subsequent elimination of water or formaldehyde (24).

Influence of toasting intensity on the presence of these compounds

Figure 16 shows the development of these aromatic molecules, extractable from toasted oak under working conditions at the cooperage. Quantitatively speaking, maltol, dihydromaltol and furaneol are the most important substances. None of these compounds are present in untoasted or lightly toasted wood. There is a sharp increase in the content of all these molecules at medium toasting, then a significant decrease after heavy toasting, apart from isomaltol, which increases at a regular rate with toasting intensity. However, this last substance, formed by the heat breakdown of maltol, has no odor. As a result, the greatest aromatic potential, giving the most intense "toasty" character, occurs after the medium toasting.

Impact of aging in new barrels on the wines' content of "toasty"-aroma volatile compounds

Table III shows the amounts of some of the previously identified molecules in the same red wine, either barrel-aged or not. Some of the samples were aged in barrels of different origins, toasted to varying degrees.

Red wine aged in stainless steel vats alone only contained slight traces of maltol, while wines aged in barrel had the various "toasty" aroma compounds identifiable in toasted wood. The abundance of these compounds varies according to the degree of toasting. The amount of furanic derivatives, considered here as a classic sign of toasting intensity, demonstrates that the quantity of enolic molecules measured in the wine correlated closely with the level of toasting.

In accordance with what we learned at the laboratory in model media, there is an optimum toasting level above which the formation of volatile substances with a "toasty" aroma increases significantly (medium toast), and a further level above which they disappear rapidly (US heavy toast for the American oak). In comparison with the standard process for European sessile oak, the toasting process developed for American oak seems more favorable towards the formation and extraction of such molecules.

Table III. Influence of aging wine in barrel on its content in enolic compounds with a « toasty » aroma (results after 6 months aging)

Volatile compounds (in μg/l)	Control		French oak		American oak			
	inox tank	[±]	medium toast	[±]	medium toast US	[±]	high toast US	[±]
2-hydroxy-3-méthyl-2-cyclopenten-1-one (cycloten)	0	0	12	1	29	2	24	2
3-hydroxy-2-methyl-4H-pyranone (maltol)	traces	0	32	2	53	4	32	2
2,3-dihydro-5-hydroxy-6-methyl-4H-pyran-4-one* (dihydromaltol)	0	0	19	1	23	2	25	2
4-hydroxy-2,5-dimethyl-3(2H)-furanone (furaneol)	0	0	12	1	12	1	13	1
2,3-dihydro-3,5-dihydroxy-6-methyl-4(H)-pyran-4-one* (DDMP)	0	0	1,5	0	2,5	0	6	0
Σ enolic compounds	0	0	76,5	5,5	119,5	8,5	100	7,2
Σ furanic aldehydes	25	1	17700	885	18540	927	26230	1312

0 : not detectable
*: in maltol equivalent

+/- : standard deviation

Concentration µg/l

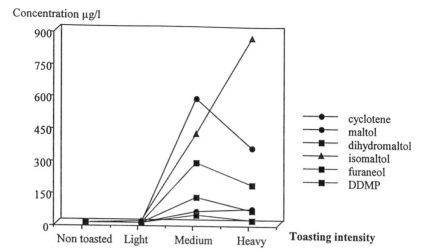

Figure 16. Development of enolic compounds during barrel toasting
(extraction by soaking in model alcohol solution 12 % v/v, 20 g/l of toasted wood, 2 weeks' contact)

Aromatic potential of molecules identified in toasted wood

Detection thresholds (α = 50 %) in the model water-alcohol solution were 5 mg/l and 2 mg/l for maltol and cycloten respectively (Table IV). DDMP has a powerful « burnt sugar » aroma at the dry state or by sniffing the effluent of gas chromatography, but it is odorless in aqueous solutions as wine. We have still not been able to measure thresholds of other molecules; DHM and furaneol could be some very interesting molecules.

In view of the concentrations which have been measured, the maltol and cycloten in the barrel wood do not seem to play an organoleptically important role in wines. It is nevertheless possible to find these same molecules in much higher concentrations in certain conditions. Some of the other molecules identified should have much lower detection thresholds and should explain the majority of the "toasty" character, typical of new barrels. Even if they have not been aged in wood, some red and white wines have a natural "toasty" aroma coming from volatile sulfur compounds as 3-methylthio-propionic acid (**8**), 2-mercapto-ethanol acetate (**9**) and 3-mercapto-propan-1-ol acetate (**10**) recently identified by Lavigne (28) in our laboratory ; 3-methythio-propionic acid is more characteristic of red wines and the two thioacetates of white wines. The specific characteristics contributed by oak thus intensify a wine's typical aromatic character. It seems, therefore, that the wood has the capacity to enhance the natural wine's « toasty bouquet ».

Table IV. Perception thresholds of « toasty » compounds from toasted oak wood and wine (in a model water-alcohol 12 % v/v solution, thresholds estimated for 50 % of the population)

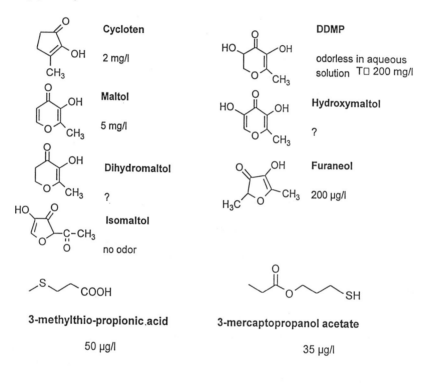

Cycloten
2 mg/l

Maltol
5 mg/l

Dihydromaltol
?

Isomaltol
no odor

DDMP
odorless in aqueous solution T□ 200 mg/l

Hydroxymaltol
?

Furaneol
200 µg/l

3-methylthio-propionic.acid

50 µg/l

3-mercaptopropanol acetate

35 µg/l

2-mercaptoethanol acetate

65 µg/l

concentration (µg/l)

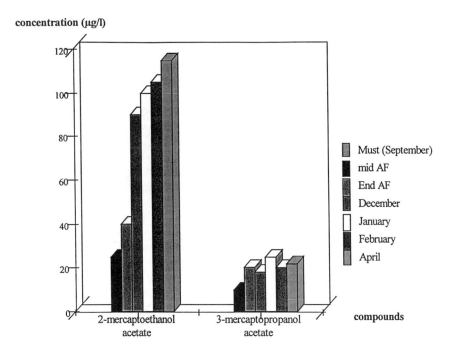

Figure 17. Changes in 2-mercaptoethanol acetate and 3-mercaptopropanol acetate during the fermentation (AF) and the aging of a dry white wine on its lees.

The monitoring of a same Semillon white wine fermented and aged in barrel during several months (figure 17) shows that the two thioacetates were not present in the must, but appeared during the alcoholic fermentation. Their respective contents rose significantly during the first months of aging. About two months after alcoholic fermentation (early December), the 3-mercaptopropanol acetate concentration stabilized, whereas the 2-mercaptoethanol content continued to increase steadily in the wine (28).

Future researches will explore synergy or complementarily between aromatic molecules of wine and toasted oak, as well as the production and aging conditions likely to bring these various types of aromas together so as to achieve the best results.

Barrel-aging has improved wine quality tremendously over the past several centuries. Guidelines for the best way to use oak are constantly evolving in order to make the best possible wines. Only full understanding of barrel-aging will enable outstanding wines to be produced.

Literature cited

1. MASUDA M., NISHIMURA K., 1971. *Phytochem.*, **10**, 1401-1402.

2. SCALBERT A., PENG S., MONTIES B., HERVE DU PENHOAT C., GAGES D., 1990. Isolation and structural characterization of polyphenols from *Quercus robur L.* and *Quercus petrae Liebl.* woods. In "Int. Meet. Groupe Polyphenols", Strasbourg, July 9-11, **15**, 203-206.

3. HERVE DU PENHOAT C.L.M, MICHON V.M.F., PENG S., VIRIOT C., SCALBERT A. et GAGE D., 1991. *J. Chem. Soc. Perkin trans.*, **1**, 1653-1660 ;

4. CHATONNET P., DUBOURDIEU D. 1997,. *Am. J. Enol. Vitic.* accepted for publication.

5. CHATONNET P. Influence des procédés de tonnellerie et des conditions d'élevage sur la composition et la qualité des vins élevés en fûts de chêne. Ph. D Thesis N°338 University of Bordeaux II.

6. CHATONNET P., DUBOURDIEU D. 1998, *J. Food Sci. Agric.* **76**, in press

7. TRESSL R., BAHRI D., SILWAR R.,1979. Bildung von Aldehyden durch lipidoxidation und deren Bedeutung als "off-flavor" Komponenten in Bier. In *"EBC Proceedings 17th Congress"*, Berlin, 27-41.

8. BARKER R.L., GRACEY D.E.F., IRWIN A.J., PIPASTS P., LEISKA E., 1983, *J. Inst Brew.*, **89**, 411-415 .

9. BADINGS H.T., 1970, *Ned. Melk-Zuiveltijdschr.*, **24**, 147-256.

10. CHATONNET P., 1991. Incidences du bois de chêne sur la composition chimique et les qualités organoleptiques des vins - Applications technologiques. Thèse de D.E.R Université de Bordeaux II, N°2, 224 p.

11. CORDIER B., CHATONNET P., SARISHIVILI N.G., OGANESYANTS L.A, 1994. *Vestnik of the Russian Academy Agricultural Science*, 3, 63-66.

12 CHATONNET P., DUBOURDIEU D., SARISHIVILI N.G., OGANESYANTS L.A et CORDIER B 1996, *Rev. Fr. Oenologie*, **97**, 167, 46-51.

13. CHATONNET P. BOIDRON J.N. et DUBOURDIEU D., 1994, *J. Int. Sci. Vigne et du Vin,* **28**, 4, 337-357.

14. GÜNTHER C. et MOSANDL A., 1987, *Z Lebensm. Unters Forsch*, **185**, 1-4.

15. GUICHARD E., FOURNIER E., MASSON G. and PUECH J.L. 1995, *Am. J. Enol. Vitic.*, **46**, 4, 419-423.

16. CHATONNET P. BOIDRON J.N. et DUBOURDIEU D., 1994, *J. Int. Sci. Vigne et du Vin,* **28**, 4, 359-380.

17. LARIGNON P., ROULLAND C., VIDAL J.P., CANTAGREL R., 1994. Etude de la maturation en charentes des bois de tonnellerie. Rapport de la Station Viticole du Bureau National Interprofessionnel du Cognac, Mars 1994, 21 p.

18. CHATONNET P. BOIDRON J.N. et DUBOURDIEU D., 1994, *J. Int. Sci. Vigne et du Vin*, **28**, 3, 185-201.

19. CUTZACH I., CHATONNET P., HENRY R., DUBOURDIEU D. 1997, *J. Agric. Food Chem.* 25, 2217-2224.

20. DUBOIS P., 1989, *Rev. Fr. Oenol.*, **120**, 19-24.

21. SEFTON M., FRANCIS I.L., WILLIAMS P.J., 1989, Volatile flavour components from oak wood, In *Proceedings of the seventh Australian Wine Technical Conference*, 13-17 August, Adelaïde, (SA).

22. MILLS F.D., HODGE J.E, 1976, *Carbohydrate Research*, **51**, 9-11.

23. HODGE J.E., FISHER B.E., NELSON E.C, 1963, *Am. Soc. Brewing Chemist Proc.* 84-95.

24. KIM M.O, BALTES W. 1996, *J. Agric. Food Chem.,* 44 282-289.

25. CHATONNET P. et BOIDRON J.N., 1989, *Conn. vigne et vin,* **23**, 1-11.

26. HODGE J.E 1967, Origin of flavors in food : nonenzimatic browning reactions. In The chemistry and physiology of flavors. 465-485, SHULTZ H.W. Ed., AVI Publishing company, Westport, Connecticut.

27. FEATHER M.S., HARRIS J.F. 1973, *Adv. Carbohydr. Chem. Biochem.* 28, 161-165.

28. LAVIGNE V. 1996 Recherches sur les composés soufrés volatils formés par la levure au cours de la vinification et de l'élevage des vins blancs secs. Ph D. Thesis N°429, University of Bordeaux II.

Chapter 15

Detection of Cork Taint in Wine Using Automated Solid-Phase MicroExtraction in Combination with GC/MS-SIM

Christian E. Butzke, Thomas J. Evans, and Susan E. Ebeler

Department of Viticulture and Enology, University of California, One Shields Avenue, Davis, CA 95616-8749

Cork taint is a musty/moldy off-odor in wine. It is related to the cork stopper, a wine bottle closure made from the bark of the cork oak (*Quercus suber*). In a correlation between sensory evaluation and chemical analysis, 2,4,6-trichloroanisole (TCA) has been identified as a major impact component. In sensitivity tests of a group of trained wine judges, a geometric mean of the minimum detectable concentrations of TCA has been determined at 4.6 ng/L.
Solid Phase MicroExtraction (SPME) is a solvent-free sample preparation method based on the adsorption of analytes directly from an aqueous sample onto a coated fused-silica fiber. Headspace SPME was used in combination with gas chromatography-mass spectrometry/selective ion monitoring (GC/MS-SIM) to analyze for TCA in wine. Wines were spiked with TCA, and its deuterated stable isotope, $^2H_5$-TCA, was used as an internal standard. The extraction fiber of the SPME, coated with polymethylsiloxane, was exposed for 25 minutes in the headspace of the sample vial, and then injected into the injection port of the GC-MS by a Varian 8200 CX autosampler. Limit of quantification of this method was 5 ng/L. The method was linear from 5 to 250 ng/L with an overall coefficient of variation for replicate analyses of less than 13%.

The wineries in the United States produce wine with a retail value of over $10 billion each year, of which they export about 3% ($326 million in 1996). Grapes are the most valuable crop in California, which produces 90% of the wine in the US, exceeded only by dairy products among all agricultural commodities produced in the state. Grape products constitute a significant and growing segment of California's agricultural exports, ranking fifth in value. Grapes rank as the ninth most valuable crop nationwide. The American wine industry imports roughly 3,600 tons of bark cork stoppers as bottle closures, worth over $80 million annually.
Corks are a major export industry for Portugal. The country produces about 78% of the roughly 23 billion bark cork stoppers used annually world wide. It manufactures ca. 70% of the world's cork products with an estimated export value of over $500 million.

Cork taint has been recognized for years as a serious off-odor problem in the wine industry. It is generally perceived as a musty, earthy and moldy aroma. At low levels in wine it causes loss in varietal fruit character and masks the aroma. Internationally, it is estimated that cork-related wine spoilage exceeds over $10 billion in value (*1*). This includes losses resulting from physical defects of corks, causing seepage, leakage and unwanted oxidation. The estimated incidence of corked wine bottles ranges from 2 to 7%, which means that wine with a retail value of $180 to 630 million just from California is being spoiled by cork taint every year. At a very conservatively estimated taint rate of 2%, the estimated costs of using cork stoppers as wine bottle closures amount to $281 million a year in the United States alone (*Table I*).

Table I: Cork taint and its economic impact on the US wine industry

Statistics	
US cork stopper imports	900,000,000/Year
Average price	$90/1000
Import value of corks	$81,000,000/Year
Est. cork taint rate	2%
Est. value of tainted wine	$200,000,000/Year
Est. costs for cork closures	$281,000,000/Year
US wine exports (1996)	$326,000,000/Year

Cork taint components

Although over 100 volatiles from finished corks have been reported (*2-3*), the one component that has been identified (*4*) as the major cause for cork taint is 2,4,6-trichloroanisole (TCA). In a recent study of Australian wines (*5*), 100% of the tainted wines, assessed by wine industry personnel, had TCA at or above the sensory threshold. The European QUERCUS study found TCA to be responsible for a musty/moldy taint in at least 80% of cases when it was detected in bottled wines. This makes TCA the most significant impact compound in regard to cork taint, and consequently made it the focus of our analytical developments.

2,4,6-Trichloroanisole

There are three major chemical/biochemical pathways through which TCA can be formed during cork production (*6*). Both involve phenols as the basic structure, a chemical chlorination step and a microbial methylation. The first, probably most important mechanism starts with the methylation of phenolic components from the cork lignin by mold growth on the cork after harvest. Among the mold genera that have been isolated from cork and identified are *Penecillium, Aspergillus, Alternaria* ("yellow stain") *Mucor , Monilia, Trichoderma, Cladosporium, Paecilomyces,* and *Rhizoctonia* (*7-9*). In addition, infections with edible fungi such as *Armillaria mellea* have been investigated (*2*).

The chlorination of anisoles present in the cork occurs during the hypochlorite wash, which has been a part of cork processing for many years, results in the formation of TCA. Equally significant is the potential methylation of chlorophenols following the hypochlorite bleaching if mold growth is not discouraged after the traditional washes. Severe contamination with mold spores will likely have occurred prior to washing or by re-contamination, e.g. in the cork processing plant. A German study (10) found 18% of hypochlorite-bleached corks to contain TCA at 6 to 13 ng/g and all to contain TCP at 19 to 301 ng/g. Unbleached corks had no detectable levels of TCA and about 2.5% were contaminated with tetrachlorophenols (TCP). However, the third source for TCA is the original presence of chlorophenols in the cork bark from environmental pollution, use of certain pesticides and herbicides in the cork forest, or absorption from wood preservatives during storage (11-13). TCPs can be methylated at any stage of processing or storage if cork moisture levels allow for mold activity.

Other polychlorinated anisoles, such as 2,3,4,6-tetrachloroanisole or the pentachloroanisoles exhibit sensory characteristics similar to TCA but have been found at much smaller levels, and mostly in combination with TCA in contaminated wines. However, both penta/tetrachloroanisoles and the corresponding chlorophenols have also been observed separately as environmental pollutants from building or shipping materials (13).

Besides TCA, a limited number of other components have been implicated in corky off-odors in wine. These are guaiacol, geosmin, 2-methylisoborneol and 1-octen-3-ol and its corresponding ketone, 1-octen-3-one (14), and possibly methoxypyrazines.

Guaiacol has a "burnt/smoky", medicinal character, and may impose a cork-derived off-odor to a wine. However, Amon's study (14) of wines characterized as "corked", could not find guaiacol concentrations above sensory threshold. Elevated guaiacol levels in wine have been associated with *Streptomyces* infections on the cork, but components such as 4-ethylguaiacol may also be produced from grape phenolic acids via vinylphenols by *Brettanomyces* yeast decarboxylase and reductase, resp. Besides, volatile phenols can be present in grape berry tissue, bound as glycosides. Methylisoborneol, found in 35% of tainted wines above threshold, displays "earthy/muddy" notes, while geosmin has "earthy" sensory properties and was found in only 14% of the tainted wines tested in the Australian study. It is unstable at wine pH with a half-life of less than eight week at pH 3.2/25°C, and its degradation products do not seem to possess the same sensory characteristics. Both geosmin and methylisoborneol are common metabolites of soil bacteria and molds which have been isolated from cork (6). Of the C_8 compounds, 1-octen-3-ol was found in 19% of corked wines and has similar "metallic/mushroomy" properties as 1-octen-3-one, which are quite distinct from moldy/musty characters. The ketone can be found in high percentages of both tainted and untainted wine, suggesting a more general contribution to wine aroma. Both components are also associated with microbial activity of cork molds and other fungi.

Elevated levels (>10 ng/L) of an isomer of a another group of extremely potent wine aroma compounds, methoxypyrazines, have been detected in some red wines (15). Isopropylmethoxypyrazine (2-methoxy-3-(1-methylethyl)pyrazine) may be associated with cork-related microbial contamination.

As mentioned earlier, even without being clearly identified as a musty/moldy off-character by the consumer, low levels of cork taint may be perceived as a general loss of a wine's fruitiness, masking its aroma.

Cork stopper production

Cork stoppers are manufactured from the outer bark (suberose parenchyma) of the cork oak, *Quercus suber L.* (16). The cork oak is grown in several Mediterranean countries, with Portugal accounting for more than half of the world-wide production of cork (ca. 170,000 t on average) and about 78% of bark wine bottle stoppers (18 billion/year). Starting out with a planted acorn, it takes, even under ideal conditions, roughly 43 years to harvest the first commercially acceptable cork. The first bark (virgin cork) is removed after about 25 years, or when the trunk girth has reached more than 0.6 m measured at 1.2 m height. After another nine years, the secondary reproduction cork bark is stripped, but it is not structurally homogeneous enough yet to produce natural (one-piece) cork stoppers. Only the third generation of bark (*amadia*) is suitable for stopper production. Its time of harvest (every 7-10 years), and therefore thickness, depends on the local growing conditions as well as the desired diameter of the stoppers to be punched out of it. Productivity in Portugal reaches on average about 2.5 t/ha cork forest.

The chemical composition of cork is made up by about 43% *suberin* (composed of fatty acids and alcohols), 28% lignin, 13% cellulose, 6% tannins, 5% waxes, and 5% ash. About 90% of the tissue is gas, resulting in a density of 0.12 to 0.20 kg/L. Cork has a unique capability as a bottle seal because of its excellent resilience after insertion into a bottle. This is due to its structure consisting of polygonal cells (30 to 42 million/cm^3) separated by spaces filled with gas (atmospheric air without CO_2) which slows oxygen diffusion without completely eliminating it.

However, gas diffusion through an intact cork stopper has been estimated at only 0.1 mg O_2/L per year (*17*), leading to one complete oxygen saturation (ca. 6 mg/L) in a bottle of wine in about 60 years. However, a wide variation in gas permeability has been reported (*18*) which can contribute to significant bottle-to-bottle variation in the detrimental oxidation of white wines. Some winemakers report anecdotal evidence about positive aging characteristics in red wines aged in cork-closed bottles presumably due to a small degree of oxygen penetration, while others do not find wine quality improvements over alternative closures such as specially lined crown caps. While white wines are normally best aged with no oxidation at all, reds can improve with at least up to 10 saturations (60 mg O_2/L) and may take up to 25 without loss in quality. Since the capacity of an individual wine to take up oxygen above its saturation level varies manifold (*17*), the perceivable effects of oxygen penetration through the cork will not be consistent if it exists at all. An increase in headspace (ullage) in older bottles is considered a sign of failure (due to eventual gas release from the filled spaces between cork cells) of a given cork as a seal, not an effect of continuous or reproducible gas exchange or evaporation through the cork.

Cork has been used as material for container closures dating back to ancient Egypt but cork stoppers as glass bottle closures have only been used for less than 300 years, replacing wooden stoppers that were hemp-wrapped and soaked in olive oil. The introduction of cork stoppers as wine bottle closures is attributed to the French Benedictine Dom Pierre Pérignon (1639-1715) who is known for his instrumental role in the evolution of sparkling wine production. In 1750, the first commercial cork stopper factory was established on the Spanish-French border.

The bark is harvested during the spring and summer months when it is growing new cells and can be more easily removed from the inner bark (phloem) covering the xylem. Opposite to oak barrel coopering, the tree is not destroyed by stripping the bark, and can be harvested for more than 100 years.

After being stripped, the cork bark boards are traditionally stored in stacks in the forest to serve as raw material for the year-round production of stoppers. The first processing step occurs when the stacked boards are boiled in large brick tubs for about one hour. The boiling softens the slightly rounded boards so they can be straightened out to ease the subsequent processing. This step also washes out some of the bitter and astringent tannins that would be undesirable to extract once the cork comes in contact with wine. A similar effect is attributed to the seasoning by rain and sun, and possibly enzymatic activity during storage in the forest, although the seasoning of wood for wine barrels has shown to result in a qualitative change in the sensory properties of oak tannins rather than in a reduction of their total concentration. The boards are air-dried while remaining stacked before they may be boiled and dried a second time. During the drying process, excessive mold growth occurs on the board, covering them with a white blanket of mycelia. In more modern facilities, the washing and drying steps are moved to the cork factory where they can be performed under controlled conditions. Autoclaving the cork boards has been recommended (*19*), since during the traditional boiling the temperature within the boards does not exceed 87°C which does not affect the viability of mold spores.

The cork boards are sliced into smaller sections (0.05 by 0.3 m) from which about 10 corks can be punched out either manually or by automated devices. Subsequently, the raw stoppers are sanded to exact size, and several manual and electronic grading steps will follow, separating the corks by visual quality only.

To sanitize and to lighten the color of the cork stoppers, they undergo several washing and/or bleaching steps. Traditionally, a chlorine bleach wash has been followed by an oxalic acid rinse. In this process, the corks are dipped into a calcium hypochlorite (30 g/L) bath for about two minutes, then held for up to two hours at room temperature before they are rinsed with water followed by an oxalic acid solution (6-8 g/L) to neutralize the oxidant. Both the resulting white deposit of calcium oxalate and the preceding bleaching improve the visual quality of the cork by simulating a more consistent surface and structure.

Due to the implication in cork taint formation, the chlorine bleaching step is being replaced by a hydrogen peroxide bleach, or completely eliminated. The hydrogen peroxide bath contain 10% H_2O_2 and 5% ammonia. The oxalic acid neutralization wash may be substituted by a citric acid (1%) rinse. Alternatively, washes in potassium metabisulfide (1%) for 5 minutes have been used, as well as sulfamic acid (2%) rinses for 10 minutes. For the treatment of the bark cork disks that are glued to an agglomerated section in the making of sparkling wine corks, ethanol and citric acid rinses at elevated temperature have been applied (*20*). Different additional processes to remove taint components or precursors from the bark have been experimented with, ranging from steam stripping to the application of ozone.

After a final rinse with clean water in tumbler washers, the cork moisture levels are adjusted to 5.5 to 8% in continuous tunnel dryers to avoid growth of microorganisms on the washed corks. An ideal moisture level achieves a water activity unsuitable for mold growth, yet high enough to maintain the flexibility of the cork necessary for bottling.

The dried corks will usually be further checked for visual quality, before being de-dusted and coated with commonly a paraffin/silicon mix (*21*) to reduce capillary effects between bottle neck and cork surface, and to ease insertion into and extraction of the cork from the wine bottle. After going through counting machines, the corks are 'branded' with a winery logo, nowadays mostly a soy-based ink imprint that is replacing the more expensive and slower hot branding. The last steps will usually be conducted at the cork supplier or at the winery in the particular wine growing region.

There are several critical processing steps the authors have observed, during which contamination with taint components or their precursors can occur. All of those potential sources for taint development need to be eliminated in order to regain winery and consumer confidence in wine bottle closures made from cork bark.

1. In or near the cork forests, any use of pesticides, fungicides or herbicides containing traces of chlorophenols must be strictly prohibited, and trees must be systematically tested for residues.
2. Since small amounts of TCA have been found on cork trees, the cork forests need to be better protected from industrial and residential air pollution, since atmospheric air will be incorporated into the cork structure.
3. Cork boards should be stripped off the tree in safe distance from the soil and cover crops should be kept as low as possible.
4. Cork boards should not be stored in the forest close to the soil with its high numbers of mold spores, and soil should not splash onto the boards during rain.

5. The boards should not be stored on treated wood of any kind, such as railroad ties or transport pallets, since those are likely to be contaminated with chlorophenols from wood preservatives or pesticides.
6. Just as cooperage wood, cork boards should be stored away from soil or spore-containing dust if air-drying and seasoning is desired.
7. Microbial growth of any kind on cork material during processing must be eliminated, not only infections that are referred to as "yellow stain".
8. The boil water must be chlorine-free and it should be exchanged frequently, since its temperature will not kill mold spores.
9. Use of chlorine bleach must be completely eliminated; the purpose of any bleaching step must be questioned. Use of chlorinated water for any wash or rinse step should be avoided.
10. Cork should not be processed or stored in locations with high air pollution, since chloro-organic compounds may be absorbed.
11. Lots of finished cork must be kept at appropriate moisture levels and completely isolated from incoming moldy cork boards.
12. Corks must be shipped and stored in moisture-controlled air-tight containers, not in open bags, in order to avoid recontamination with mold, excessive moisture levels, and absorption of ubiquitous TCPs or TCA from the storage environment.

Corks should be considered a part of the wine and treated like a food product throughout the process. Only if at least those basic criteria are met, sources of musty/moldy off-odors other than cork stoppers may be considered. Moldy barrels, fining agents, packaging and building materials, or freight containers contaminated with TCA and TCPs, as well as the excessive use of chlorinated sanitizing agents have been named, but at an industry-wide cork taint rate of 2 to 7 percent do not appear to play any significant role in the overall problem.

Sensory thresholds

Suprenant (22) found that individual thresholds for TCA can vary substantially. Although group thresholds can be calculated, they only give a very rough estimate of the minimum amounts of taint perceived by wine consumers The thresholds also vary as a function of the specific wine that is tasted. Sensitivity to TCA can be improved by training, but insensitivity (anosmia) may also occur in individuals. Sensitivity is decreasing with age and and is dramatically lower in smokers. A group threshold (geometric mean) for TCA (in a neutral Sauvignon blanc wine) of 46.6 ng/L has been been reported, while for trained judges in a sensitivity test (minimum detectable concentration, MCD), a geometric MCD mean of 4.6 ng/L was found (23). Meilgaard (24) found the average order of magnitude of variation in individual thresholds to be less than one (10 times). While a variation of two orders of magnitude for published threshold values is not uncommon in the literature, those may be considered artifacts due to impurities in the compounds tested or the use of untrained judges. The concentrations of TCA found in tainted wine range from 22 to 374 ng/L.

Cork sensory quality control

Although some efforts have been made (25-26), until now there have been no statistically sufficient, rapid and cost effective sensory quality control (QC) procedures against cork taint for either wineries or cork suppliers. At UC Davis, we have developed a Cork Sensory QC Manual (27) which evaluates the major valid sampling procedures, and exemplifies their use based on a case study at a premium winery in California. In addition, the manual provides instructions for a taste panel evaluation which is essential for wineries in order test their winemakers' and cellarmasters' as well as the tasting room staff's sensitivities to detect the off-odor. Once individual sensory thresholds for TCA are determined, the manual can be used to establish proper training procedures to set-up a winery taste panel for cork sensory QC. The sensory evaluation of large sample numbers required to achieve "acceptable" cork taint rates of below 1%, takes substantial personnel involvement and logistics. The combination of correct sampling procedures and a new automated instrumental analysis that is capable of testing large sample sizes at human sensory threshold levels will allow us to better quantify the extent and distribution patterns of cork taint, eventually protecting wineries from shipments of badly tainted corks.

Cork taint analysis

Since the human sensory threshold for cork taint is in the low parts-per-trillion range, the instrumental analysis for the impact compound TCA is especially challenging. Previously reported assays all involved labor intensive liquid-liquid extractions, using significant amounts of solvents, in combination with gas chromatography (GC). Buser et al. (28) employed a pentane/ethyl acetate (3:1) solution as the extracting solvent in combination with an adsorption onto a silica gel minicolumn. In spiked TCA additions ranging from 30 to 100 ng/L, recoveries of 43 to 72% have been reported with quantification via an external standard procedure. Whitfield et al. (29) applied a similar technique for the analysis of 2,3,4,6-tetrachloroanisole, pentachloroanisole and TCA in packaging material and in dried fruit. They used 3,5-dimethyl-2,4,6-trichloroanisole as the internal standard (IS), and obtained recoveries of 82 to 105% for the three analytes in the different sample matrices. In fiberboard samples, coefficients of variation (CV) ranged from 5 to 8%. Both liquid/liquid extraction procedures reported limits of quantification (LOQ) for TCA of 2 to 5 ng/L. Furthermore, trichloroethane extraction followed by a distillation step has been utilized by Spadone for TCA determination in coffee (30), and direct thermal desorption from a cork sample with cryo-focusing on the GC column was used by Hoffmann (31). Sefton's group (5) analyzed TCA in wine samples by extraction with n-pentane followed by fractional distillation and concentration in a stream of nitrogen. In samples spiked with TCA at levels of 2 ng/L, this group reported a mean value of 2.2 ng/L over six replicates with a CV of 18%.

Solid Phase MicroExtraction (SPME)

SPME is a patented sample preparation method for GC applications (32-36). The solvent-free technique was developed in 1989 by Janusz Pawliszyn (*http://www.science.uwaterloo.ca/~janusz/spme.html*) at the University of Waterloo in Ontario, Canada, and a manual device made by Supelco, Inc. has been available since 1993. In 1996, Varian Associates, Inc., constructed the first SPME autosampler. SPME involves exposing a fused silica fiber that has been coated with a non-volatile polymer to a sample or its headspace. The absorbed analytes are thermally desorbed in the injector of a gas chromatograph for separation and quantification. The fiber is mounted in a syringe-like holder which protects the fiber during storage and penetration of septa on the sample vial and in the GC injector. This device is operated like an ordinary GC syringe for sampling and injection. The extraction principle can be described as an equilibrium process in which the analyte partitions between the fiber and the aqueous phase.

We have used headspace SPME in combination with GC/mass spectrometry-select ion monitoring to analyze for cork taint impact compounds in wine. *Table II* lists the analytical parameters that the authors developed for the analysis of TCA in wine (37).

Limit of quantification of this method was 5 ng/L. The method was linear from 5 to 250 ng/L. The accuracy (8%) and precision (CV below 13%) we observed with this protocol using the SPME procedure are very similar to those obtained with the traditional liquid-liquid extraction assays.

Table II: Analytical parameters for TCA analysis by SPME-GC/MS-SIM

Parameter	
Gas chromatograph	HP-5890
Mass selective detector	HP-5971/72
Autosampler	Varian 8200 CX
Column	25m x 0.25mm x 0.25μm, non-polar CP-SIL 5 CB
SPME fiber	100μm polydimethylsiloxane coat (Supelco, Inc.)
SPME time	25 min absorption at 45°C, 3 min desorption at 260°C
Samples	10 mL liquid per 16 mL vial; headspace sampling
Internal standard	deuterated $^2H_5$-TCA
Injector	260°C, splitless
Temperature program	45°C/2 min - 12°C/min - 265°C/1 min
SIM fragments	m/z 195 (TCA); m/z 215 ($^2H_5$-TCA)

Cork research and future of bark cork stoppers

The causes for the majority of cases of cork taint have been quite thoroughly investigated mostly due to efforts from Australia (38-40) and to a certain degree from Europe. Although the problem of cork taint poses enormous financial losses upon wineries worldwide, the funding of research efforts in different wine producing regions has been quite variable (*Table III*).

Table III: Estimated cork taint research funding 1992-1997

Region	Funding (US$)
Europe	2,400,000+
Australia	100,000+
USA	13,000

Only if all cork producers improve their often antiquated processing procedures and equipment as well as their shipping methods, so that corks are eventually processed and handled like a food product, will the cork stopper have a future as a closure for wine bottles. Being a natural and therefore renewable and biodegradable product, cork possesses unique properties among packaging materials that are worth preserving. As a wine bottle closure, cork stoppers have been an integral part of the traditional wine drinking experience, and their charmingly unpredictable removal poses an intriguing challenge every time a bottle is opened. With the technical changes and quality control techniques outlined in this article being implemented by both cork suppliers and wineries, the wine consumers should enjoy un*corking* their favorite and untainted beverage for many generations to come.

Acknowledgments

The authors most gratefully acknowledge Dr. Mark Sefton, Australian Wine Research Institute, Adelaide, for supplying us with deuterated TCA as internal standard. We would also like to thank ETS Laboratories, St. Helena, California, and Varian Associates, Walnut Creek, California, for their support.

Literature Cited

1) Fuller, P. *Aust. NZ. Wine Ind. J.* **1995**, 10(1): 58-60.
2) Rocha, S.; Delgadillo, I; Correia, A. J. *J. Ag. Food Chem.* **1996**, *44(3),* 865-71.
3) Mazzoleni, V.; Caldentey, P.; Careri, M.; Mangia, A.; Colagrande, O. *Am. J. Enol. Vitic.* **1994**, *45(4),* 401-406.
4) Tanner, H.; Zanier, C.; Buser, H. *Schw. Z. Obst Weinbau* **1981**, *117,* 97-103.

5) Pollnitz, A.P.; Pardon, K.H.; Liacopoulos, D.; Skouroumounis, G.K.; Sefton, M.A. *Aust. J. Grape Wine Res.* **1996**, *2*, 184-190.
6) Lee, T.H.; Simpson, R.F. In *Wine Microbiology and Biotechnology*; G.N Fleet, Ed.; Harwood Academic Publishers, Philadelphia, PA, 1993, 353-372.
7) Codina, J; Esteban, C.; Calvo, A.; Agut, M., *Industrie delle Bevande* **1993**, *22 (128)*, 561-563.
8) Daly, N.M.; Lee, T.H.; Fleet, G.H. *Food Tech. Australia* **1984**, *36(1)*, 22-24.
9) Heimann, W.V.; Rapp, A.; Volter, I.; Knipser ,W. *Deutsche Lebensmittel-Rundschau* **1983**, *79*, 103-107.
10) Sponholz, W. R.; Muno, H. *Industrie delle Bevande* **1994**, *23(130)*, 133-138.
11) Cooper, J.F.; Tourte, J.; Gros, P. *Chromatographia* **1994**, *38(3/4)*, 147-150.
12) Rigaud, J.; Issanchou, S.; Sarris, J.; Langlois, D. *Sciences des Aliments* **1984**, *4*, 81-93.
13) Chatonnet, P.; Guimberteau G.; Dubourdieu D.; Boidron, J.N. *J. Int. Sci. Vigne Vin.* **1994**, *28(2)*, 131-151.
14) Amon, J.M.; Vandepeer, J.M.; Simpson, R.F. *Aust. NZ. Wine Ind. J.* **1989**, *4(1)*, 62-69.
15) Allen, M.S.; Lacey, M.J.; Boyd, S.J. *J. Ag. Food Chem.* **1995**, *43(3)*, 769-772.
16) De Oliviera, M.A.; de Oliviera, L. In: *The Cork*, Grupo Amorim, Portugal, ISBN 972-95525-0-9, 1991.
17) Singleton, V.L. In *Principles and Practices of Winemaking*, Chapman & Hall, New York, NY, 1995, 410-422.
18) Waters, E.J.; Peng, Z.; Pocock, K.F.; Williams, P.J. *Aust. J. Grape Wine Res.* **1996**, 2, 191-197.
19) Rocha, S.; Delgadillo, I.; Ferrer-Correia, A.J. *J. Ag. Food Chem.* **1996**, *44(3)*, 872-876
20) Neel, D. *Pract. Winery Vyd.* **1993**, *14(4)*, 11-14.
21) Fugelsang, K.; Callaway, D.; Toland, T.; Muller, C.J. *Pract. Winery Vyd.* **1997**, *17(5)*, 50-55.
22) Suprenant, A.; Butzke, C.E. In *Proc. Intern. Symp. Cool Climate Vitic.*, Am Soc. Enol.Vitic., Rochester, NY, 1996
23) Suprenant, A., *M.S. Thesis*, Dept. of Vit. and Enol., University of California, Davis, CA, 1997
24) Meilgaard, M.C. *J. Am. Soc. Brew. Chem.* **1991**, *49(3)*, 128-135.
25) Cavazza, A; de Ros, G.; Zini, C., *Enotecnico* **1995**, 31 (12) 65-72.
26) Panaiotis, F.; Tribaut-Sohier, I.; Valade, M. *Deutsche Weinbau* **1995**, *18*, 14-16.
27) Suprenant, A.; Butzke, C.E. *Cork Sensory Quality Control Manual*, DANR, University of California, Davis, CA, 1998.
28) Buser, H.R.; Zanier, C.; Tanner, H. *J. Ag. Food. Chem.* 1982, 30: 359-362.
29) Whitfield, F.B.; Shaw, K.J.; Nguyen, T.H. *J. Sci. Food. Ag.* **1986**, *37*, 85-96.
30) Spadone, J.C.; Takeoka, G.; Liardon, R. *J. Agric. Food. Chem.* **1990**, *38(1)*, 226-233.
31) Hoffmann, A; Sponholz, W.R. In: *Proc. 16th Intern. Symp. on Capillary Chromatography*, Riva del Garda, Italy, 1994
32) Arthur, C.L.; Pratt, K.; Motlagh, S., Pawliszyn, J. *J. High Res. Chrom.* **1992**, *15*, 741-744.
33) Louch, D.; Motlagh, S.; Pawliszyn, J. *Anal.Chem.* **1992**, *64*, 1187-1199.
34) Bucholz, K.D.; Pawliszyn, J. *Anal. Chem.* **1994**, *66*, 160-167.
35) Yang, X.; Peppard, J. *J. Ag. Food Chem.* **1994**, *42*, 1925-1930.
36) Yang, X.; Peppard, J. *LC-GC* **1995**, *13(10)*, 882-886.
37) Evans T.J.; Butzke, C.E.; Ebeler, S.E. *J. Chrom. A.* **1997**, *786*, 293-298.
38) Amon, J.M.; Simpson, R.F. *Aust. Grapegrower Winemaker* **1986**, 268, 63-80.
39) Simpson, R.F. *Aust. NZ. Wine Ind. J.* **1990**, *5*, 286-296.
40) Leske, P. *Aust. NZ. Wine Ind. J.* **1996**, *11(1)*, 36-39.

Chapter 16

Flavor–Matrix Interactions in Wine

A. Voilley[1] and S. Lubbers[2]

[1]Laboratoire de Génie des Procédés Alimentaires et Biotechnologiques, Université de Bourgogne, ENSBANA, 1 place Erasme 21000 Dijon, France
[2]Laboratoire de Biochimie Alimentaire, ENESAD, 26 Boulevard Petitjean 21000 Dijon, France

The interactions between aroma compounds and other components of a wine matrix : colloids, fining agents and ethanol were investigated in model systems and with instrumental methods. The physico-chemical interactions between aroma compounds and other components depend on the nature of volatile compounds. The level of binding generally increases as the hydrophobic nature of the aroma increases. The interactions also depend on the nature of the macromolecules such as yeast walls, mannoproteins, bentonite or smaller molecules such as ethanol. As a function of the nature of non-volatile component, the increase or decrease in the volatility of aroma compounds can influence largely the overall aroma of wine. The effect of ethanol on the volatility of aroma compounds is understood and it clearly appears that ethanol leads to modification in macromolecule conformation such as protein, which changes the binding capacity of the macromolecule. This review enables to develop some hypotheses on the possible sensory contribution of some non-volatile compounds of wine on the overall aroma.

Food flavor is a very important parameter influencing perceived quality. The volatile compounds contributing to the aroma of foods possess different chemical characteristics, such as boiling points and solubilities and the sensory properties of food cannot be understood only from the knowledge of aroma composition. This can be explained by interactions between flavor compounds and major constituents in food such as fat, proteins and carbohydrates (1). A number of different interactions has been proposed to explain the association of flavor compound with other food components. This includes reversible Van der Waals interactions and hydrogen bonds, hydrophobic interactions. The understanding of interactions of flavor with food is becoming important for the formulation of new foods or to preserve flavor characteristics during processing. As in the case of other food products, flavor characteristics of wine or spirits cannot be understood only from the knowledge of their composition. Interactions between aroma compounds and other non-volatile components are likely to play a role in quality of wine.

Quality is a perception that is not easy to describe, especially in wine. Some of the more positive obvious factors of wine are the distinctive aromas derived from some grape varieties (2). Recognizable modifications produced by viticultural practices, climate, winemaking style, processing and aging also may be highly regarded. When too accentuated or unfamiliar these same features may be considered as defects. Most defects modify the subtle characteristics which distinguish the wines from one to another. Balance and harmony in wine commonly refers to the sapid (taste and mouth feel) and olfactory (flavor) sensations.

From chemical point of view, wine is an acidic aqueous ethanol solution with aroma compounds. Organic acids, colloids, polyphenols and mineral salts are 2% of the wine composition. However winemaking is complex and some minor components are likely to play an important role in sensory properties of wine. Carbohydrates and glycoprotein colloids in wine, deriving from grape and yeast, constitute a small amount of the dry matter of wine. In white wine, the concentration is 150-400 mg/L (3-4). Winemaking processes, such as the aging of white wine on lees, lead to an increase of colloids. This processing is considered as an important factor of quality of Burgundy white wines. The special characteristic of the wines is attributed to components released during the autolysis of yeast. The concentration of these colloids increases, especially glucans and mannans from yeast cell walls (5-6). In contrast, processing for clarification and preserving wine leads to an elimination of colloid material. Fining agents such as caseinate, bentonite or membrane cross-flow filtration can greatly modify the sensory character of wine in particular conditions (7).

Further high quality wines are traditionally matured in oak cashs for several months. Maturation reactions are complex and like in spirit beverages, dissolution of wood components is of prime importance (8-9-10). Extracted wood components have secondary effect other than their direct flavor contribution. They appear to be necessary for correct maturation of the beverage. The effect of wood components on the maturation of beverages was clearly investigate in distilled alcoholic beverages only. The addition of oak extract to a model spirit solution reduced the extractability of ethyl esters with chain lengths of 10-20 carbons by the dichloromethane (9-10). The increase in solubility reflects a reduction in the activity of the ester in the aqueous ethanol solution, relative to the dichloromethane and indictes that an interaction between components of the wood extract and esters takes place.

At last the sensory part of ethanol in wine is important , it plays a major role on the volatility of flavors and in the interactions between aroma compounds and other components.

This paper reviews the interactions between aroma compounds and other components of a wine matrix : colloids, fining agents and ethanol. Studies are carried out with model systems and instrumental methods to investigate flavor-matrix interactions.

Methods of Studing Interactions

Model Wine and Aroma compounds. The model wine was composed of an aqueous solution of ethanol 125 mL/L, L(+) malic acid 3 g/L, acetic acid 0.106 mL/L, K_2SO_4 0.1 g/L, $MgSO_4$ 0.025 g/L. The pH of the model wine was adjusted to 3.5 with NaOH. Macromolecules were added at 1 to 10 g/L to the model wine.

Aroma compounds selected were : isoamyl alcohol (100µL/L), isoamyl acetate (100µL/L), ethyl hexanoate (100µL/L), ethyl octanoate (40µL/L), ethyl decanoate (10µL/L), octanal (100µL/L), β-ionone (100µL/L), γ-decalactone (100µL/L), supplied by Aldrich (Steinheim, Germany). They are all slightly soluble in water except for isoamyl alcohol which is soluble in water. The hydrophobic constants of volatile compounds are expressed by Log P where P is the partition coefficient of the compound in water/octanol system.

Activity Coefficients of Volatile Compounds. The headspace technique was used to determine the activity coefficients of volatile compounds as described previously *(11)*. The headspace system flask contained 10 to 20 mL of the model wine with the diluted volatile compound, at 25°C. The flow rate of nitrogen gas in the flask was 5 to 10 mL/min. The concentration in volatile compound in the vapor phase was analysed by gas chromatography. The conditions were reported in a previous paper *(11)*. The relative volatility of the volatile compound can be expressed as a partition coefficient K and activity coefficient γ.

Partition coefficient :
$$K^\infty = \frac{y}{x}$$

Activity coefficient :
$$\gamma^\infty = K^\infty \frac{P_t}{P_i^s}$$

P_i^s vapor pressure of the pure volatile compound at 25 °C (mm Hg) ; P_t total pressure (mm Hg) ; x mole fraction of the volatile compound in the solution ; y mole fraction of the volatile compound in the vapor phase.

Equilibrium Dialysis Method. The equilibrium dialysis method is based on the diffusion of the volatile compound through a semi-permeable membrane placed between two compartments containing the model wine and macromolecules *(11)*. In the experiment, 1mL solution of macromolecule in the model wine was placed on one side of the membrane (compartment 1) and 1 mL of the model wine containing a known amount of the volatile compound on the other side (compartment 2). The system was shaken at 30 °C for 12 h to reach equilibrium of the free ligand (volatile compound) between the two compartments of the cell. At equilibrium, the concentration of the volatile compound was determined by gas chromatography. The difference in concentration of the volatile compound between the two compartments represents the amount of the volatile compound bound to the macromolecule.

Interactions between Yeast Derived Non-Volatiles and Aroma Compounds

The interactions between aroma compounds and macromolecules from yeast released during alcoholic fermentation (F) and autolysis (A) were studied by the headspace technique *(11)*. The values of infinite dilution activity coefficients of volatile compounds were measured in a model wine with and without macromolecules at 1g/L (Table I). The volatility of ethyl decanoate stays constant in the presence of both extracts. For ethyl hexanoate and octanal, the F extract produces a significant (P< 0.01) decrease in the activity coefficient, by 12 and 8% respectively. Conversely F extract increases the volatility of isoamyl alcohol and ethyl octanoate by 6 and 19% respectively. The A extract increases the volatility of ethyl hexanoate by 6% and the volatility of ethyl octanoate by 15%. These results demonstrate the complex influence of macromolecules from yeast released during fermentation or autolysis on the volatility of aroma compounds.

Table I. Values of activity coefficient in the model wine without extract and with fermentation extract (F) and autolysis extract (A) at 1 g/L. Value in brackets is the standard deviation.

Aroma compound	Model wine	F extract	A extract
Isoamyl alcohol	61 (1)	65 (1)*	63 (1)
Octanol	6117 (91)	5644 (56)✦	6200 (90)
Ethyl hexanoate	9424 (77)	8282 (58)✦	8800 (75)*
Ethyl octanoate	300233 (1050)	358414 (2513)✦	340900 (2565)✦
Ethyl decanoate	3775000 (37277)	3775000 (37260)	3585000 (34520)*

Values are significantly different at * P< 0.05, ✦ P< 0.01.
Adapted from ref. 11.

These extracts are a mixture of glucans and mannoproteins from yeast cell walls. The effects of each component of the extract on the interaction with aroma substances are complex. Therefore the purification of protein and mannoproteins populations from F extract was undertaken to investigate the nature of macromolecule which can explained the binding of aroma compounds. The interactions of β-ionone and ethyl hexanoate with the different fractions obtained from the F extract were studied by the equilibrium dialysis method. Four fractions were obtained by ion exchange chromatography from F extract. The first fraction obtained by IEC, named F1, represents 42% of the F extract. This fraction was chromatographed on concanavaline A Sepharose. Affinity chromatography on Con-A allows isolation of glucans and mannoproteins which have low affinity to Con-A (fraction N) and on the other hand, mannoproteins having high affinity to Con-A (fraction R). β-ionone is significantly bound on the macromolecules of F1 fraction at 7.5% (Figure 1). The macromolecules of N-F1, with a great proportion

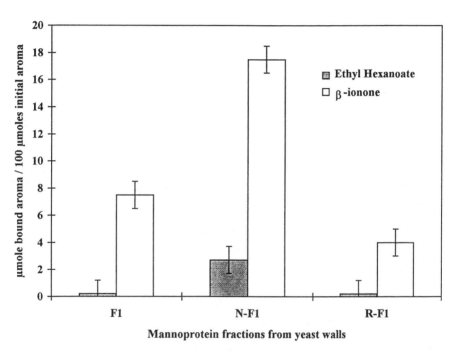

Figure 1. Percentage of binding with F extract and the fractionsN-F1 and R-F1 (from affinity chromatography on Con-A) at 10 g/L in model wine by the equilibrium dialysis method.

of proteins (60 g/100g dry matter), isolated by affinity chromatography from F1, bind 17.5% of β-ionone and 2.7% of ethyl hexanoate. The extend of binding of β-ionone between the F1 fraction and the most purified fraction N-F1 increases by a factor of 2.5. The highly glycosylated mannoproteins (R-F1) which constitute the main component of the total extract bind β-ionone weakly (4%) and do not interact with ethyl hexanoate. The polysaccharide part constituted 90% of these macromolecules. The polysaccharide chains in mannoproteins are composed of polymannose. This mannan structure does not contain inclusion site as do starch *(12)* ; this can interpret the low binding efficiency for aroma compounds. Therefore the level of binding depends largely on the amount of protein in the mannoproteins.

This study demonstrated the influence of natural colloids from wine (mannoproteins released from yeast) on the volatility of aroma compounds and therefore the possible role of these minor components of a wine matrix on sensory properties of wine. The physico-chemical interactions between aroma substances and exocellular yeast material depend on the nature of volatile compounds and of the macromolecules.

Interactions between Proteinaceous and Fining Non-Volatiles and Aroma Compounds

Several treatment agents of wine : yeast cell walls, sodium caseinate, gelatin, bentonite were evaluated for their potential to bind with aroma compounds. The loss of sensory properties of wine, especially flavor modification, is partly caused by protein stabilization treatments with fining agents or ultrafiltration processing of wine *(13-14)*. Yeast cell walls are used in sluggish or stuck wine fermentation ; the effect on fermentation has been explained by the adsorption of toxic fatty acids present in the growth medium *(15)*. Therefore yeast walls are also assumed to bind aroma compounds.

Yeast Cell Walls. Interactions between aroma substances and yeast walls induce to a modification of the volatility of some aroma compounds in the model wine *(16)*.Yeast walls do not bind a specific chemical class of volatile compounds (Table II). The volatility of octanal, an aldehyde and of ethyl hexanoate, an ester, decreases by 14% with yeast walls at 1 g/L. The effect of walls is greater on the volatility of ethyl octanoate than that of the other aroma compounds ; the partition coefficient decreases by 45% for ethyl octanoate in the presence of 1 g/L yeast cell walls.

Table II. Percentage decrease in headspace concentration of aroma compound with yeast walls at 1 g/L in model wine. Log P is the hydrophobicity constant.

Aroma compound	log P	% decrease in headspace concentration
Isoamyl alcohol	1.21	9
Octanol	2.64	14
Ethyl hexanoate	2.76	14
Ethyl octanoate	3.88	45

Adapted from ref. 16

The hydrophobic nature of the volatile substance seemed an important factor. The volatile compound with the highest hydrophobic constant (log P = 3.88), ethyl octanoate, is bound to a larger extent on yeast walls. Conversely, isoamyl alcohol, with log P= 1.21, is less fixed ; the decrease in volatility is 9%.

The presence of lipid fraction in yeast cell walls explains partly the binding of aroma compounds. The high amount of lipids in the industrial yeast cell walls resulted from the manufacturing process. The yeast walls were obtained after autolysis of whole cells. The plasma membrane was destroyed and lipids were able to be adsorbed onto the yeast wall surface. Yeast walls free of lipids were studied by equilibrium dialysis method with β-ionone and ethyl hexanoate (Figure 2). It was found that lipid-free yeast walls bound volatile compounds to a lower extent. The greater decrease in the binding on yeast walls for β-ionone compared to ethyl hexanoate can be attributed to the higher lipid solubility of β-ionone. However, lipid-free yeast walls always bind some volatile compounds, i.e. 22% for ethyl hexanoate and 50% for β-ionone. Therefore the binding capacity of yeast walls is not only due to lipid matter, mannoproteins also play a role.

Yeast cell walls are present in white wines which were aged on lees. Therefore we can suggest that yeast walls from lees influence on the equilibrium of the bouquet of the wine.

Fining Agents. The binding capacity of caseinate, used for fining white and red wines was measured by heaspace analysis (Table III). Sodium caseinate at 1 g/L in model wine decreases the volatility of β–ionone more than that of ethyl hexanoate and isoamyl acetate. Like yeast walls, the most hydrophobic compound is the most bound to a larger extent.

Table III. Percentage decrease in the activity coefficients of aroma compounds in model wine containing fining agent at 1 g/L of proteins

Aroma compound	log P hydrophobicity constant	Sodium caseinate
Ethyl hexanoate	2.76	25
Isoamyl acetate	2.12	6
β-ionone	4.14	49

An other fining agent is used in white must and wine : bentonite. In a model system, bentonites showed important differences on the binding capacity of the aroma compounds, i.e. γ-decalactone and β-ionone (17). The binding capacity of bentonite is not negligible, therefore some bentonites bind up to 25% of the aroma compounds present in the solution (Table IV). The binding of the aroma compounds on two bentonites were measured in model must (model wine without ethanol + glucose and fructose at 100 g/L), in model wine and in must and wine of

224

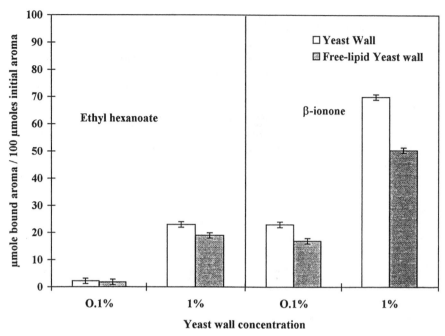

Figure 2. Percentage of binding with yeast cell walls and lipid-free yeast walls at 1 g/L and 10 g/L in the model wine by the equilibrium dialysis method.

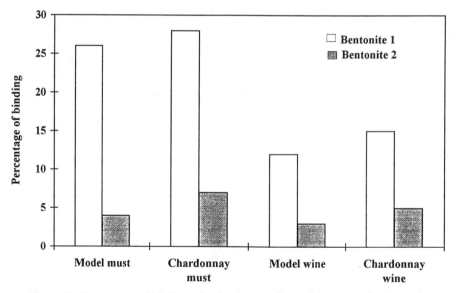

Figure 3. Percentage of binding of γ-decalactone in model must and wine and in must and wine of Chardonnay.

Chardonnay (Figure 3). The binding of γ-decalactone on bentonite increases of 6 fold in the presence of monosaccharides (model must and must of Chardonnay) compared to the control. Therefore the loss of aroma compounds caused by bentonite fining is higher in must than in wine. However weak effects on the behaviour of the flavor of wine were suggested. Indeed many of the aroma compounds in wine are produced during the alcoholic fermentation. Therefore these interactions may have an important effect on the flavor properties of the finished wine.

In model wine (hydroalcoholic solution), binding of γ-decalactone on bentonite is lower than in blanck solution. No effect of ethanol has been shown. Results obtained from model solutions were closed with results obtained from must and wine of Chardonnay.

Table IV. Effect of bentonites (1 g/L) on removal protein and binding of aroma compounds

Bentonite coded name	Removal protein %	γ-decalactone % bound	β-ionone % bound
LA1	73	17	25
IM1	77	23	23
LO2	45	0	0
OF2	82	3	9
MV2	78	9	16
MV4	23	0	0

Adapted from ref. 17

Importance of Ethanol in Wine-Flavor Interactions

Effect of Ethanol on Volatility of Aroma Compounds. The activity coefficients of volatile compounds obtained by headspace method are lower in the presence of ethanol at 126 ml/L than in water (Table V). The headspace responses of aroma compounds are reduced by one-half (11-18). The aroma compounds are not very polar and are more soluble in ethanol than in water ; hence the activity coefficient decreases, as shown by other authors for alcoholic beverages (19). This effect of solubilisation can be explained by the presence of interactions between aroma compound, water and ethanol.

Effect of Ethanol on Conformational state of protein. To understand the effect of ethanol and pH in flavour-protein interactions the binding of γ-decalactone to bovine serum albumin was investigated using the equilibrium dialysis method (20). Without ethanol, a decrease in pH (from 5.3 to 3.5) reduces by one-half the γ-decalactone binding onto protein. In the presence of ethanol, changing pH do not have any appreciable effect (Table VI).

Ethanol appeared to modify flavor binding phenomena and it seemed that ethanol affected the conformational state of proteins. The relationship between the surface hydrophobicity of protein which informs on the conformational state and the

Table V. Values of activity coefficients (γ) of aroma compounds in water and model wine. Value in brackets is the standard deviation.

Aroma compound	Water	Model wine
Isoamyl alcohol	107 (1) ✦	61 (1) ✦
Octanal	12230 (100) ✦	6117 (91)✦
Ethyl hexanoate	18950 (102) ✦	9424 (77) ✦
Ethyl octanoate	599466 (1120) ✦	300233 (1050) ✦

Values are significantly different at ✦ P< 0.01.
Adapted from ref. 11

Table VI. Percentage of binding of γ-decalactone on BSA at 10g/L

pH	Without ethanol	With ethanol 10% w/w
3.5	17	13
5.3	34	14

Adapted from ref. 20

binding of aroma compounds in the presence of ethanol was investigated. A fluorescent probe (1-anilino-8-naphtalene sulfonic acid) was used in the surface hydrophobicity determination (21-22). Bovine serum albumin (BSA), ovalbumin and trypsin inhibitor were studied at 10 g/L in a model wine with and without ethanol in presence of three aroma compounds. Ovalbumin and trypsin inhibitor bind significantly less β-ionone than BSA (Table VII). In the presence of ethanol, aroma compounds are less bound onto these proteins. The binding of aroma compound onto BSA decreases by a factor of 2, 1.09 and 1.37, respectively with ethyl hexanoate, γ-decalactone and β-ionone. With ovalbumin, the binding of β-ionone decreases of 2 fold. Therefore ethanol seems to be predominant in the decrease of the flavor binding onto proteins. Druaux et al. (20) showed a decrease of 4.8 fold in the binding affinity of BSA for γ-decalactone in the presence of ethanol, also suggesting conformational changes in protein.

Table VII. Molar percentage of bound aroma compounds on proteins in model wine with or without ethanol

	Bovine serum albumin		Ovalbumin		Trypsin inhibitor	
	No ethanol	Ethanol	No ethanol	Ethanol	No ethanol	Ethanol
Ethyl hexanoate	16.4	8.2	3	1	3.6	2.6
β-ionone	35.8	32.7	13.8	6.8	18.3	7.2
γ-decalactone	17.9	13.0	1.2	1	3.3	2.3

Adapted from ref. 21

The study of the surface hydrophobicity of protein in the presence of ethanol confirmed the above result. The apparent dissociation constants for BSA and ovalbumin increased from 1.5 to 1.9×10^{-6} M and 6.8 to 7.9×10^{-6} M respectively in the presence of ethanol in citrate buffer *(21)*, while the number of binding sites decreased from 22 to 10 for BSA and from 40 to 19 for ovalbumin. Therefore it clearly appears that ethanol leads to modifications in protein conformation which causes changes in surface hydrophobicity. This result is consistent with the decrease in the binding of aroma compounds to proteins in the presence of ethanol.

Conclusion

The knowledge of the composition of volatile compounds in food has greatly increased during the past decade. Many studies continue to report the identity and the concentration of volatile compounds in food matrices. However concentration alone appears insufficient to explain flavor properties of food. The lack of our knowledge concerning the influence of non-volatile constituents of food on the perception of aroma has to be filled by studies such as those presented in this paper. Data on interactions between aroma and matrix in wine are scarce compared with other food matrices studied. Flavor-matrix interactions in wine have generally been obtained in model systems and with instrumental experiments. However it is possible to develop some hypotheses on the possible sensory contribution of some non-volatile compounds of wine on overall aroma.

The physico-chemical interactions between aroma compounds and other components depend on the nature of volatile compounds. The level of binding generally increased with the hydrophobic nature of the aroma. However interactions also depend on the nature of macromolecules such as yeast walls, mannoproteins, bentonite or smaller molecule such as ethanol. As a function of the nature of non-volatile component, the increase or decrease in the volatility of aroma compounds can influence largely the overall aroma of wine.

In natural colloids from wine, the binding can be attributed to mannoproteins containing a high proportion of proteins. In treatment agents all the products can binding aroma compounds. Fining agents with protein origin such as caseinate and with mineral origin such as bentonite can bind great amounts of aroma compounds.

The alteration of flavors of wine observed when colloids were largely eliminated can be explained by the retention of aroma compounds on the macromolecules eliminated by fining processes. On the other hand, the absence of these macromolecules which increase the aroma intensity of some volatile compounds such as ethyl octanoate could incidence general modifications in the flavor equilibrium of wine.

The effect of ethanol on the volatility of aroma compounds is shown and it clearly appears that ethanol leads to a modification in macromolecule conformation such as protein, which changes the binding capacity of the macromolecule.

In the future, the effect of wood components on the flavor of wine has to be investigated. The interactions between wood components and some esters have been shown in spirit model solution. These studies in spirits suggest that the maturation in wood of wine could influence the distribution of aroma compounds between more than two phases (ethanol and water).

So it is quite clear that we still have a long way to more understand the behaviour of wine flavors during winemaking and aging.

Acknowledgments

Works performed under financial support from Conseil Régional de Bourgogne, Ministère de l'Enseignement Supérieur et de la Recherche and Fould Springer Industry.

References

1. Bakker J. In *Ingredients Interactions;* Gaonkar, A.G.; Ed.; Marcel Dekker : New York, US, **1995**; pp 411-439.
2. Moio, L.; Chambellan, E.; Lesschaeve I.; Issanchou, S.; Schlich, P.; Etiévant P. *J. Food Sci.* **1995**, *3*, 265-278.
3. Llauberes, R.; Dubourdieu, D.; Villetaz, J.C. *J. Sci. Food Agric.* **1987**, *41*, 277-286.
4. Saulnier, L.; Mercereau, T.; Vezinhet, F. *J. Sci. Food Agric.* **1991**, *54*, 275-286.
5. Feuillat, M.; Freyssinet, M.; Charpentier, C. *Vitis*, **1989**, *28*, 161-176.
6. Ferrari, G.; Feuillat, M. *Vitis* **1988**, *27*, 183-197.
7. Feuillat, M.; Peyron, D.; Berger, J.L. *Bull. O.I.V.* **1987**, 663-674.
8. Boidron, J.N.; Chatonnet, P.; Pons, M. *Conn. Vigne et Vins* **1988**, *22*, 275-294.
9. Pigott, J.R.; Conner, J.M.; Clyne, J.; Paterson, A. *J. Sci. Food Agric.* **1992**, *59*, 477-482.
10. Conners, J.M.; Paterson, A.; Pigott, J.R. *J. Sci. Food Agric.* **1994**, *66*, 45-33.
11. Lubbers, S.; Voilley, A.; Feuillat, M.; Charpentier, C. *Lebensm. Wiss. Technol.* **1994**, *27*, 108-114.
12. Rutschmann, M.A.; Heiniger, J.; Pliska, V.; Solms, J. *Lebensm. Wiss. Technol.* **1990**, *22*, 240-244.
13. Miller, G.C.; Amon, J.M.; Gilson, R.L.; Simpson, R.F. *Australi. Grape Winema.* **1985**, *256*, 49-50.
14. Voilley, A.; Lamer, C.; Dubois, P.; Feuillat, M. *J. Agric. Food Chem.* **1990**, *38*, 248-251.
15. Alexandre, H.; Lubbers, S.; Charpentier, C. *Food Biotechnol.* **1997**, *11*, 89-99.
16. Lubbers, S.; Voilley, A.; Charpentier, C.; Feuillat, M. *Am. J. Enol Vitic.* **1994**, *45*, 29-33.

17. Lubbers, S.; Charpentier, C.; Feuillat, M. *Vitis* **1993,** *35*, 59-62.
18. Kepner, R.E.; Marse, H.; Strating, J. *Anal. Chem.* **1964**, *36*, 77-82.
19. Bakker, J.; Brown, W.; Hills, B.; Boudaud, N.; Wilson, C.; Harrison, H. In *Flavour Science*; Taylor, A.J.; Mottran, D.S.; Ed.; The Royal Society of Chemistry, London, GB, **1996**; pp 369-374.
20. Druaux, C.; Lubbers, S.; Charpentier, C.; Voilley, A. *Food Chem.* **1995,** *53*, 203-207.
21. Lubbers, S.; Voilley, A.; Charpentier, C.; Feuillat, M. in *Bioflavour 95*; Etiévant, P.; Schreier, P.; *Ed.;* Les colloques 75; INRA editions: Paris, Fr, **1995**; pp127-129.
22. Langourieux, S. *Thesis Ph.D.; Université de MontpellierII, France* **1993**.

INDEXES

Author Index

Subject Index

A

Acetal(s), effect on wine, 169

Acetaldehyde
characterization and measurement in wine, 166–178
role in red wine flavor, 124–139

Acetaldehyde-induced polymerization, role in wine flavor, 125

Acid, role in bitterness and astringency, 160–162

Acid-catalyzed reactions, biomimetic synthesis of solerone, 116–112

Acid hydrolysates, role in aroma of Cabernet Sauvignon and Merlot wines, 13–28

Acid-labile glycoconjugates of monoterpenoids and C_{13}-norisoprenoids in Riesling wine, role as flavor precursors, 1, 36

Activated carbon adsorbents, use in fining, 143–144

Activity coefficient, definition, 219

Agar, use in fining, 143

Aging of oak wood, role in volatile and odoriferous compounds, 187, 189–192

Aging of wine, role in red wine flavor, 124–139

β-Alanine, role in hydrogen sulfide production during wine fermentations, 81–94

Aldehyde–anthocyanin condensation reactions, effect on wine, 169, 171f

Aldehyde dehydrogenase, effect on aldehydes in wine, 168

Aldehyde(s) in wine
chemical reactions in grapes and wines, 169, 171f
comparison of MS and nitrogen phosphorus detectors, 172–173, 174t
derivatization pH vs. recovery, 173
experimental description, 168
experimental materials, 170
experimental procedure
derivatization and extraction, 172
formation in grapes and wines, 168
GC, 172
standard curve, 172
future work, 176, 178
measurement in grapes and wine, 169–170, 171f
recoveries, 173, 174t
role in aging characteristics and color stability, 166
role in flavor, 166, 167t
SO_2 addition effect
during fermentation, 173, 176, 177t
prior to bottling and storage, 176, 177t

Aldehyde–tannin condensation reactions, effect on wine, 169, 171f

Amino acids, role in hydrogen sulfide production during wine fermentations, 81–94

Analysis, labile terpenoid aroma precursors, 1–10

Antioxidant activity, role of small-scale fining in Merlot wine, 142–153

Aroma, role of glycosidic precursors in Cabernet Sauvignon and Merlot wines, 13–28

Aroma compound interactions
with fining agents, 223–225
with yeast cell walls, 220, 222–223, 224f

Aroma extract dilution analysis, odor profiles of white wine varieties, 39–51

Aroma of *Vitis vinifera* L. cv. Scheurebe, role of volatile compounds, 53–63

Aroma precursors, analysis, structure, and reactivity, 1–10

Astringency
definition, 156
hypothesis for mechanism, 164
mechanism, 157
red wines, 156
role
acid, 160–161